Spatial Health Inequalities

Adapting GIS Tools and Data Analysis

Spatial Health Inequalities

Adapting GIS Tools and Data Analysis

Esra Ozdenerol

CRC Press
Taylor & Francis Group
Boca Raton London New York

CRC Press is an imprint of the
Taylor & Francis Group, an **informa** business

CRC Press
Taylor & Francis Group
6000 Broken Sound Parkway NW, Suite 300
Boca Raton, FL 33487-2742

First issued in paperback 2019

© 2017 by Esra Ozdenerol
CRC Press is an imprint of Taylor & Francis Group, an Informa business

No claim to original U.S. Government works

ISBN-13: 978-1-4987-0150-1 (hbk)

ISBN-13: 978-0-367-87144-4 (pbk)

Visit the Taylor & Francis Web site at
http://www.taylorandfrancis.com

and the CRC Press Web site at
http://www.crcpress.com

Contents

major discussion of regional disparity that involves the north, the south, the rural, and the urban, in addition to the rich and the poor.

The approach used in this book is clever and useful. This book demonstrates the spatial health inequalities in six most important topics in environmental and public health, including food insecurity, birth health outcomes, infectious diseases, children's lead poisoning, chronic diseases, and health-care access. These are the topics that the author has done extensive research on. Each chapter addresses a topic. In each chapter, this book will provide a detailed description of the topic from a global perspective, identify relevant data and data sources, discuss key literature on appropriate techniques, and then illustrate with real data with mapping and GIS techniques. It is important to note that while spatial analysis is useful, inappropriate applications of the spatial analytical methods and a lack of understanding of the errors and uncertainties involved in spatial analysis could generate erroneous results and interpretations. This book will discuss the appropriate techniques and bring awareness of the possible errors and uncertainties involved in spatial analysis. Finally, this book has a web-based resource that contains data from the case studies relevant to each health topic. The web page provides links to download data and instructional material to supplement the case studies.

Spatial health inequalities are intertwined with issues in environmental justice and community resilience; they are all closely interrelated. GIS methods are the prime tools to reveal the patterns, identify the factors, assess the progress, and ultimately plan for better health resilience. This book is a timely and valuable addition to the literature. It will be an effective text and resource for those who want to understand the issue of spatial health inequalities and how to analyze the issue with GIS and spatial analysis.

Nina Lam, PhD
Professor and E.L. Abraham Distinguished Professor
of Louisiana Environmental Studies
Department of Environmental Sciences
Louisiana State University
Baton Rouge, Louisiana

Foreword

It is my great pleasure to write a foreword for this book, written by a talented medical geographer. Since the turn of the century, there have been quite a few books on spatial epidemiology, statistical and geographic information system (GIS) approaches in medical geography, and spatial analysis for health and environment studies. The uniqueness of this book is its focus on spatial health inequalities, which are pervasive in the United States as well as many countries in the world. This book is the first to put together GIS, mapping, and spatial analytical methods to illustrate how these various methods can be used to analyze spatial health inequalities.

Health inequality or disparity has long been a global health problem. Health inequality, as this book states, refers to different health outcomes of individuals due to their differences in socioeconomic status. Low-income, underserved minority groups are more vulnerable to diseases and epidemics and are more likely to lead to adverse health outcomes. In other words, borrowing the concept of resilience in disaster studies, individuals that have higher socioeconomic statuses are generally more resilient to adverse health effects; they generally have better prevention, are more prepared, and bounce back faster than individuals that have lower socioeconomic statuses.

Extending health inequality to spatial health inequality represents a major conceptual as well as practical advance in the study of epidemiology and environmental health. As this book states, health inequality involves both the person and the place. The neighborhoods and the biophysical, political, and cultural environments all play key roles in affecting the health outcomes of individuals. Unequal spatial distribution of resources such as clinics, hospitals, public transportation, fresh food markets, and schools could make the community as a whole more vulnerable and less resilient to adverse health effects, leading to lower community resilience of health. Thus, analysis of health inequalities must also consider its spatial context or larger environment to better understand the problem.

GIS and spatial analytical methods are most apt in studying spatial health inequalities. Moreover, traditional statistical or data mining methods when enhanced with GIS mapping and analysis can help reveal more effectively the spatial patterns of disparity and the factors associated with the patterns. There are many examples in the literature, as well as shown in this book, on how GIS and mapping can demonstrate spatial health inequalities and help in deriving better and more useful hypotheses. For example, my previous studies on mapping China's cancer mortalities by county in the early 1970s revealed high concentrations of different types of cancers at different localities. Through an intense data collection and mapping effort on AIDS incidence and mortality in the United States in the 1980s, we were able to confirm a national trend of AIDS spreading to rural America, generating a

This book is dedicated to my two sons, Derin and Deniz; my dad, the Honorable Justice Erdogan Ozdenerol; my mother, Umran Ozdenerol; my sister, Emel Ozbay; my brother-in-law, Alp Ozbay; my two beautiful nieces, Irem and Zeynep Ozbay; and my deceased grandmother, Sidika Erolozden. My family has given me endless support and undivided affection over the years even though there were oceans between us. Thank you for all you have sacrificed during this process.

Preface

My interest in writing this book stems from the quest to investigate the belief "who you are depends upon where you are." Having visited many countries in the Developing World and living in the Developed World, through eye-opening, perspective-building cultural experiences, I have often wondered why certain places had such stark differences in particular health outcomes. As visitors we can partake in the most popular tourist activities in diverse places such as the United States, Turkey, or India or Peru; many of us carefully crop off our photos as part of a skillful weave to craft images and often induce jealousy. We crop out the misery and isolate ourselves from the host environment's vulnerable populations affected by inequalities. This cropped-out misery could show extreme poverty conditions faced by the bitterly poor and document the miserable, hardscrabble existence of those who struggle to meet life's most basic needs. The neighborhoods and the biophysical, political, and cultural environments all play a key role in affecting health outcomes of individuals. Unequal spatial distribution of resources such as clinics, hospitals, public transportation, fresh food markets, and schools could make some communities as a whole more vulnerable and less resilient to adverse health effects. This somber reality suggests that it is rather the question of "who you are depends upon where you are" and the fact that health inequality is both a people and a place concern. That is why health inequality needs to be investigated in a spatial setting to deepen our understanding of why and how some geographical areas experience poorer health than others.

As its title *Spatial Health Inequalities* implies, the goal of my book is to introduce how spatial (i.e., geographic) context shapes health inequalities. This spatial perspective enhances the role that geographic information systems (GISs) can serve in understanding health inequalities. This understanding is critical to research and decision-making with regard to health inequalities in a global society. There is no existing text, however, helping health professionals to select a body of knowledge to help answer specific research questions related to GIS and spatial analysis to address particular health inequalities. It seems that the phrase "spatial health inequality" hasn't really permeated into common parlance. I hope that my book, *Spatial Health Inequalities*, will help remedy this situation.

This is a unique textbook for students, geographers, clinicians, health and research professionals, and community members interested in applying GIS and spatial analysis to the study of health inequalities. At the University of Memphis, I teach an upper division undergraduate and a graduate level course titled "GIS and Human Health," but I have seen similar-sounding courses at other universities with names like "Geohealth," "Geospatial Health,"

"Medical Geography," "Spatial Epidemiology," or "Spatial Analysis in Epidemiology." All of these courses aim at the audience for which *Spatial Health Inequalities* was written.

The diverse nature of various health inequalities makes it impossible for one book to cover each health inequality topic and how they are best understood and addressed with GIS and spatial analysis. However, with my book, I have focused on a set of health outcomes that are widespread and at the same time common cases of populations experiencing health inequalities in the world. My examples make use of data mostly from the Southern United States but also include examples beyond the United States. The results of this book can serve as a valuable resource for setting relevant research agendas around the world.

Chapter 1 provides an overview of the current role of GIS in the context of strategic health planning. The patient-centric and population-oriented uses of GIS are reviewed. Disease and bioterrorism surveillance, crisis and disaster management, and strategies for use of interactive maps are discussed. Conclusions present a research agenda toward a GIS-supported health information technology and how we can move beyond the map to view and explore health inequalities.

Chapter 2 discusses the use of GIS in chronic disease investigations. GIS analysis can help with monitoring and tracking chronic disease trends; identifying, implementing, and evaluating effective interventions; and providing guidance for chronic disease management programs. A representative set of chronic diseases such as diabetes, obesity, and cancer are covered. Data resources, data limitations, and the factor of geographic scale for obesity are discussed. GIS's role in cancer health disparities is summarized. As Chapter 2 points out, the important issue is how our improved understanding of the chronic disease process leads to better prevention and intervention and improves access to health services.

Chapter 3 introduces data sets related to birth morbidity and highlights issues that affect data integration in GIS with detailed discussion of the spatial patterns and characteristics of birth health such as low birth weight, birth defects, and infant mortality. From these spatial patterns, meaning may be derived about their underlying mechanisms, including individual and local environmental risk factors and access to health-care services.

Chapter 4 details the breadth of GIS applications and approaches used to show the applicability of spatial analysis for vector-borne and nonvectored diseases and future implications. Specifically, the spatial epidemiology of two vector-borne diseases, West Nile virus (WNV) and Lyme disease (LD), and their regional variations are explored. This chapter also discusses how spatial concepts and GIS-based tools have been used in understanding waterborne diseases, airborne diseases, sexually transmitted diseases, and drug addiction.

Chapter 5 focuses on current gaps in children's blood lead screening strategies and also on methods that have been most effective in examining spatial epidemiology of childhood lead exposure. This chapter also discusses additional methods of GIS utilization in research on lead poisoning and how GIS methods provide public health practitioners and policy makers the tools to better target lead poisoning preventive interventions.

Chapter 6 describes GIS's role in determining disparities in health-care access, the diverse health data sets used in small-area health-care variation studies, and data limitations. The final sections examine spatial issues in rural and regional disparities, distribution of health professionals, and evaluation of the effectiveness of intervention measures.

Chapter 7 is a conclusion chapter that deals with the developments and evolution of GIS in the context of health disparities. Future research needs, new technology initiatives, data confidentiality issues, and international collaborative efforts toward eliminating health disparities are discussed.

A notable feature of *Spatial Health Inequalities* is its companion web-based resource that provides case studies relevant to each health disparity topic covered in the text. The case studies provide hands-on application and training in the concepts covered in each health disparity chapter. This web page provides links to download data and instructional material to supplement case studies. Each case study also provides a set of links to free Internet resources to help students further explore the world of health inequalities and get directly involved with some of the GIS and mapping applications. There are a lot of materials out there on the Internet, ranging from interactive mapmaking to real-time risk mapping, and these links introduce students to it.

Any errors in this book are my own. I would appreciate hearing from discerning readers about errors of commission or omission so that they may be addressed in future editions.

As the Turkish saying goes, patience is bitter, but its fruit is sweet. I hope that I have conveyed the amazing breadth of GIS utility in studying health inequalities. I believe this book has the potential to create an extraordinary fertile new field "spatial health inequalities," one which has significant potential for real-world benefit through better understanding and management of health inequalities.

Esra Ozdenerol
Memphis, TN

Acknowledgments

I owe a great deal to the many people who have provided inspiration, help, and support for what would eventually become this book:

Irma Britton, senior editor, and Melisa Sedler, project coordinator, of CRC Press for patience and guidance throughout this entire project. Thanks to the Shelby County, Tennessee, Health Department Office of Epidemiology and Infectious Diseases for the provision of some of the maps used in my examples. I am indebted to epidemiologists David Sweat, Jennifer Kmet, Tamal Chakraverty, Sulaiman Aizezi, Dr. Lilian Ogari, and Chaitra Subramanya for collegial research partnership over the years. Thanks also to Mia Madison, Frank Burhart, and Mairi Albertson from the city of Memphis Division of Housing and Community Development for the provision of Food Deserts map. I thank my students, Cem Akkus, Su Yang Kang, Ela Bialkowska-Jeliska, and Thomas Campbell, who contributed to the development of the maps and formatting.

I thank my department head, Dr. Daniel Larsen, for giving me release time to write the manuscript and administrative assistant Julia Crutchfield for her contagious smile. I owe my perpetual gratitude to Professor Nina Lam, Abraham Distinguished Professor of Louisiana Environmental Studies, who wrote the foreword to this book.

I also thank all my friends who have lent their support and listened to my complaints for the past years. Thanks to Ashley Fowler, Johanna Leibe Gamard, Bunny Williams, Leanne Crenshaw Ince, Lisa Marie Holt, Amy Weinstein, Nazan Gursakal, Ozge Kovarik, Asli Yalcin, and Funda Cam. I thank Campus School Principal Dr. Susan Copeland and teachers, Undria Cage, Ashley Arnold, Logan Caldwell, Aeron Cassidy, LaTonya Faulkner, Lisa Miller, Jennifer Smith, Cathy Bailey, Ginny Hillhouse, Valeria Walters, Jennifer Hughes, Susan Van Dyck, Diane Coleman, and Clerisa Smith and office assistants Jackson and Higgins. Their commitment and dedication to my sons' education made my life easier but motivated me to be a better parent and an educator. I also thank James Bingham and Todd Kaplan for their immeasurable help during stressful times and for proving to me that some integrity still exists in the legal profession.

Finally, I owe a debt of gratitude to Dr. Gregory Taff, Dr. Roy Van Arsdale, Dr. Jill Nault, and Cathy Pantik who reviewed the various stages of the manuscript. Thank you for your insightful and constructive comments, which have helped to shape the final product. I much appreciate Dr. Gregory Taff's efforts in reading the manuscript, identifying errors, and suggesting improvements. This book is dedicated to my sons and my parents for their continued love.

Finally, I thank all the enthusiastic supporters of geographic information system (GIS). After teaching GIS for more than 15 years, I know that you—like my own students–will enjoy this subject. Go to GI-Yes!

Author

Esra Ozdenerol has been an associate professor in the Department of Earth Sciences of the University of Memphis since 2003. She is also affiliated with the Departments of Preventive Medicine and Health Outcome Policy of the University of Tennessee Health Science Center, Memphis, Tennessee. She is the director of the GIS Certificate Program at the University of Memphis. She also directs the Spatial Analysis and Geographic Education Laboratory in the Department of Earth Sciences. Dr. Ozdenerol was the associate director of Benjamin L. Hooks Institute for Social Change of the University of Memphis from 2010 to 2013. She obtained her doctorate in geography in 2000 and her master of landscape architecture in 1996 from the Louisiana State University. She received her bachelor of science in landscape architecture and agricultural engineering from the University of Ankara, Turkey. Before joining the University of Memphis, she was an assistant professor of architecture of the Florida International University in Miami from 2000 to 2003. Dr. Ozdenerol specializes in geographic information systems and has served as a technical consultant to various public, government, and international agencies. Her current research interests entail the use of the geospatial technologies (including geographic information, remote sensing, and cartographic and geostatistical analyses) in a diverse range of public and environmental health issues. Her latest publications involve studies about spatial health inequalities. In her spare time, she writes children books and travels. She enjoys cooking and spending time with her sons, Derin and Deniz, and dog, Ingi. She also loves to play sports, be active, and work out.

Introduction

Definition of Spatial Health Inequalities

Health inequality is a remarkably spatial phenomenon and a global health problem. A central theme is the enduring character of health inequality in the world, consistently affecting particular populations and geographies over time. What is it about certain places that have such stark differences in environment; access to, utilization of, and quality of health care; health status, or a particular health outcome? By knowing where more needs to be done, can we ensure that all persons in the world have a more equal chance to achieve their best health? Or in simple terms, how can we alleviate health inequalities?

Health inequality refers to differences, variations, and disparities in the health achievements of individuals and groups. The individual approach views health inequality as the outcome of differences of how individuals make health choices, whereas group differences (i.e., socioeconomic, gender, and race) in health could reflect the unequal distribution of resources (e.g., access to education, health care, jobs), environmental factors, or cultural factors affecting health. Though determining which inequalities stem from injustice or what proportion of an inequality is unjust is a concern of the field of health inequity.

The term and title of this text *Spatial Health Inequalities* come from the fact that health inequality is both a people and a place concern. There is a growing interest in documenting the health effects of variables that tell us things about the places (defined as neighborhoods, regions, and countries) or contexts (political, cultural, and environmental) producing health inequalities. In other words, it matters not simply "who you are in relation to where you are," but rather the question of "who you are depends upon where you are" (Kawachi et al. 2002). That is why health inequality needs to be investigated in a spatial setting. Furthermore, the characteristics of places and the people that inhabit them are dynamic in time, as well as space. The processes and interactions occurring between people and places over time are very important for health as well. This spatial and spatial–temporal perspective enhances the role that geographic information systems (GISs) can serve in both understanding and managing health inequality. People adopting GIS for tackling health inequalities need to understand GIS methods and data analysis. This book is organized to foster this understanding. It involves mapping disease data and then striving to explain the spatial distribution.

This book is written for geographers, public health practitioners, epidemiologists, students, health and research professionals, and community members interested in applying GIS and spatial analysis to the study of health inequalities. The main questions I seek to answer for the reader is, How do I analyze health inequalities using GIS? and, Which spatial measures do I use to assess health inequalities? This book follows a template, which consists of chapters, unique to each health inequality topic (i.e., food insecurity, birth health outcomes, infectious diseases, children's lead poisoning, chronic diseases, health-care access) that discusses data and data collection processes, key literature for the appropriate spatial analytical methods applied, and illustrations of cartographic techniques for visualizing and mapping health inequalities. This book lays out a step-by-step approach to people with an interest in health inequalities in their geographic context and draws examples from the United States and the rest of the world to identify relevant applications.

Factors Contributing to Health Inequalities

In this globalizing world, the importance of biological, behavioral, environmental, social, and cultural factors in driving or sustaining health inequalities is increasingly recognized. Since time and distance are compressed worldwide, these factors that contribute to health inequalities interconnect and overlap over particular geographies. Examining how these factors influence the health of individuals, groups, and populations deepen our understanding of why and how some geographical areas experience poorer health than others.

Social determinants of health, where we live, our income, our educational and employment status, and our social relationships, help determine our health status. The risk factors of ill health, many noncommunicable diseases, accidents, injuries, and violence, are found most often in the most marginalized and low-income populations (Murray and Lopez 1997). Increases in social status are paralleled by increases in health. There is a robust association between higher levels of education and lower risks of ill health. Lack of health insurance, increased health risks from lifestyles and impoverished environments, and barriers to access to quality health care are related to health inequalities associated with education (Iniguez and Palinkas 2003). Inequalities in health care are a clear contributor to inequalities in health outcomes. Differences in utilization patterns and quality of care indicators between health disparity populations and the general population have been well documented (Andersen and Newman 2005).

Rural populations without access to comprehensive health services suffer from both short- and long-term effects on their quality of life (Weeks et al. 2004). Concurrently, the increasing concentration of the world's poor in urban settings also presents higher levels of ill health (Zerre and McIntyre 2003). Job security has been shown to have an effect on life expectancy (Evans 2001).

Social context and policies represent a broad set of health determinants, encompassing health-care systems, access to health care, educational and labor policies, political power structure such as absence of democracy, household authority, workplace policies in the public or private sector, legislative authority, and cultural assets such as privileged lifestyles, high-status consumption practices, social networks, and social ties (Grusky 1994). Public and private sector policies have great potential to influence health inequalities, in both positive and negative directions. The governments' approach to domestic violence, substance abuse, and industry regulations such as transportation are among the spectrum of policies that have a bearing on health inequalities. Local, state, or national policies not specifically related to health, including those related to zoning or housing, government entitlements, public education, immigration, and criminal justice can increase or reduce health inequalities.

Beyond social determinants, characteristics present at birth such as gender, race, and ethnicity are important dimensions along which health inequalities could be described. Health and social systems may not fairly accommodate biologically specific health needs of men and women. There are differences in mortality or survival between men and women, and these differences vary by country. Differences in the roles societies accord males and females stratify their opportunities for good health. Eliminating health inequalities among racial and ethnic groups is complex, because of the complicated nature of interactions between barriers, service use, mediators of care, and the interactive impact of those factors on health outcomes (Shi 2007).

Environmental factors put populations at risk for health inequalities (Lee 2002). Racial/ethnic minority neighborhoods and tribal communities confronting neighborhood stressors such as residential segregation, poverty, and neighborhood deprivation and pollution sources are disproportionally affected by increased rates of morbidity, mortality, and adverse health outcomes (Payne-Sturges et al. 2006). Pollution of air, water, and soil along with the reduction of nonrenewable natural resources, disposal of toxic waste, global warming, ozone depletion, emerging and remerging infectious diseases, and new environmental hazards are common environmental determinants of health inequalities. Unhealthy production and consumption patterns in globalizing economy are associated with new environmental hazards and parallel with emerging and remerging infectious diseases. Low-income countries are still challenged with providing clean water and sanitation to their poor (Evans 2001). Industrial exposures to pollutants among high-risk communities, and the intensity and duration of exposure to unfavorable environments prove that early life environment sets individuals onto life trajectories that in turn affects health status over time and adversely affects health status (Kuh and Ben-Shlomo 2004). Therefore, life course perspectives across the dimension of time are important elements toward understanding health inequalities (Hertzman 1999).

How these factors can be addressed to achieve meaningful and lasting changes in health and health behavior of health disparity populations needs more research. Using individual-level instead of population-level measures will rarely be adequate for properly detecting socioeconomic and environmental gradients in health (Amick 1995). The underlying geography of these risk factors and the spatial perspective gained by GIS toward understanding health inequalities at the population level are not always emphasized in traditional public health curricula or research. This book fills this gap and provides a range of spatial methods to tackle health inequalities more effectively.

Vulnerable Populations Affected by Inequalities

Vulnerability refers to susceptibility to illness. The definition proposed by Aday states that

> Vulnerable populations are diverse groups of individuals who are at greater risk of poor physical, psychological, and/or social health.
>
> **Aday (2001)**

In the physical domain, vulnerable populations include those with physical needs such as high-risk mothers and infants, the chronically ill and disabled, and persons living with HIV/AIDS. Chronic conditions include diabetes, hypertension, dyslipidemia, heart disease, and respiratory diseases (Aday 2001). In the psychological domain, those with chronic mental conditions including bipolar disorder, major depression, schizophrenia, and attention-deficit/hyperactivity disorder and those who are suicidal or prone to homelessness, as well as those with a history of alcohol and/or substance abuse are considered vulnerable populations. In the social domain, vulnerable populations include those living in abusive families, the homeless, immigrants, and refugees.

From a global perspective, WHO defines vulnerability as "the degree to which a population, individual or organization is unable to anticipate, cope with, resist and recover from the impacts of disasters" (Wisner and Adams 2002). When a disaster strikes, children, pregnant women, elderly people, malnourished people, and people who are ill or immunocompromised take a relatively high share of the disease burden. Poverty, malnutrition, homelessness, poor housing, and destitution contribute to vulnerability. There are substantial differences in access to material resources such as clean water, sanitation, and adequate nutrition and housing because of income. The WHO is pursuing efforts to eradicate specific diseases (tuberculosis [TB], malaria, HIV/AIDS, childhood diseases such as polio) that heavily affect low-income countries and improve access to drugs and vaccines for the poor. The WHO Millennium Development Goals for developing countries call for significant reductions in poverty and improvements in population health (Annan 2000).

The political theorist Henry Shue also describes any population deprived of "basic rights" should be regarded, for the purposes of medical research, as a vulnerable population (Shue 1996). He considers the rights to goods like food and basic health care as primary, because the enjoyment of such basic rights is necessary for other rights such as the right to free speech and freedom of choice to exist. Special justification should be required for medical research involving participants from vulnerable populations in developing countries, just as it is for prisoners and other groups in developed countries (Zion et al. 2000).

Although there is a partial overlap between vulnerable populations and populations at risk, in concept they are different. The former are defined by shared social characteristics, whereas the latter are characterized by homogeneously high level of exposure to a single or multiple risk factors (Frohlich and Potvin 2008). Vulnerable populations usually exhibit worse health outcomes than others do, with fewer resources to improve their conditions and typically face greater barriers to accessing timely and needed health care (Lurie 1997).

A vulnerable population in the context of health disparities was broadly defined by U.S. Congress as "health disparity population" if there is "a significant disparity in the overall rate of disease incidence, prevalence, morbidity, mortality, or survival rates in the population as compared with the health status of the general population" (U.S. Congress 2004). Disparities include a difference in access, quality, and availability of care (Campanelli 2003). This broad definition considers characteristics other than race/ethnicity, such as high-risk mothers and children, elderly people, socioeconomically disadvantaged and rural populations, the chronically ill and disabled, the mentally ill, persons with AIDS, alcohol and substance abusers, veterans, and individuals with living conditions, that pose special challenges to health-care delivery (e.g., homeless, institutionalized, or homebound patients, the suicide- or homicide-prone , abusing families, nonnative speakers of the home country, non-English-speaking recent immigrants and refugees) (Kilbourne et al. 2006). Attention to disparities in health between racial and ethnic minority populations was first documented in 1985 in the Report of the Secretary's Task Force on Black and Minority Health (Heckler 1985). This document revealed that racial and ethnic minority populations suffered disproportionately from heart disease, stroke, cirrhosis, diabetes, infant mortality, unintentional injuries, and homicide. Although scientific and technological discoveries have improved the health of the U.S. population overall since then, Blacks/African-Americans, Hispanics/Latinos, American-Indians/Alaska Natives, Asian-Americans, Native Hawaiians, and other Pacific Islanders, socioeconomically disadvantaged populations, and rural populations continue to experience a disproportionate burden of disease.

Increased morbidity, mortality, incidence of disease, disability, and adverse outcomes in cancer, cardiovascular disease, diabetes, HIV/AIDS, infant mortality, and other conditions are well documented in the recent reports and the scientific literature (ODPHP 2000). Vulnerable populations experience significant disparities in life expectancy, access to and use of health-care services, morbidity, and mortality. Their health needs intersect with social and economic conditions they experience. The condition of vulnerable population is exacerbated by existing social and economic risk factors. For example, individuals living in poverty are much more likely to be in poor health and less likely to have used health care (CDC 2013). Accurate knowledge of the health insurance status of vulnerable populations is of critical importance for targeted policy interventions (Shi 2000). The numbers of these vulnerable populations are increasing, not only as the ranks of the uninsured grow, but as the population ages. Adding one or more chronic health conditions to this increasing population alarms policy makers on how to deal with the demands of this population on systems of care.

Legacy Cases and Effectiveness of Regulatory Interventions

The continuing legacy of poor health in minority populations in American society and segregated racial/ethnic minority groups in other countries is one compelling reason to take a closer look at the role of discrimination, cultural differences in lifestyle patterns, inherited health risks, and social inequalities that are reflected in discrepancies in access to health care. The health effects of discrimination, residential segregation, and social economic inequality are well documented (Mays et al. 2007). Obtaining race and ethnicity data is vital to develop and implement effective prevention, intervention, and treatment programs and enforceable standards to ensure nondiscrimination; facilitate the provision of culturally and linguistically appropriate health care; and identify and track similarities and differences in performance and quality of care in various geographic, cultural, and ethnic communities.

Although the United States has a population more diverse than any other nation in the world, current federal practice and policies do not fully reflect the legal foundation supportive of collection and reporting of racial, ethnic, and primary language data that are very critical in reducing disparities

(Lurie et al. 2005; 2006). The Office for Civil Rights and the Surgeon General issued the "Aetna letter," a policy directed principally at health plans, reaffirming the legality of racial and ethnic data collection (Perot and Youdelman 2001). The Department of Health and Human Services (HHS) Plan to Improve the Collection and Use of Racial and Ethnic Data, issued in 1999, provided an important impetus for policy implementation (Institute of Medicine 2009). The Operational Policy Letter No. 93 (OPL 99.093) advised health plans participating in the Medicare Choice program that race and ethnicity information may be collected on a voluntary basis from beneficiaries before, during, or after enrollment (Perot and Youdelman 2001). Collection and reporting efforts have challenges such as inconsistent or conflicting policy messages, fears of potential misuse or misinterpretation of data, lack of enforcement, lack of uniform standards for data collection, lack of a centralized authority governing data collection, or technical difficulties with data collection and maintenance. Citing the use of language data as a proxy for identifying one's immigration status or racial and ethnic data limiting enrollment in managed care plans are examples of ways in which this information could be employed for discriminatory purposes (Perot and Youdelman 2001). Upgrading technological infrastructure and redesign of work flow and data processes require significant investments.

Despite such challenges, a growing number of health plans participating in the National Health Plan Collaborative (NHPC) have detected disparities among commercially insured and Medicaid populations by collecting data from either self-reported or estimated methods on their enrolled members (Lurie et al. 2008). Proxy methodologies were used to predict race/ethnicity of their members. One of those methodologies utilizes enrollee's first and last names as proxies. First names are matched to an internally developed list of minority first names. Surnames are matched to a series of surname list from the U.S. Census.

Another methodology uses residential address as a proxy. Where a patient lives can itself have a large impact on the level and quality of health care the patient receives. Location matters in the measurement and interpretation of health and health-care disparities. In the United States, African-American and Hispanic populations tend to live in different areas from non-Hispanic white populations. GIS is used to derive indirect race and ethnicity information for health plan members. For example, GIS's geocoding tool determines the physical location of a health plan enrollee's address that has the benefit of allowing members to be tied to census data such as education levels, language proficiency, and income. These data are linked together to determine the probability that a person is Asian, African-American, Hispanic, or white/other. NHPC uses interactive mapping and analysis tools as a "geographic Pareto chart" to identify areas for interventions and make comparisons by region and race/ethnicity (Tufte 2001).

Health-care organizations also assess whether there is a "business case" or a positive return on investment in a given intervention. They also like to clarify "social case," whether the interventions provide a benefit to the patient or to the population in terms of their health status (Lurie et al. 2008). The social case could be quantified in terms of increased productivity from better health and reductions in absenteeism or prevented costs for disability payments (Dall et al. 2004).

For accurately determining the business case, routine collection and use of race/ethnicity data are needed along with strong evaluation designs that could isolate intervention effects. In the absence of rigorous research evaluation, the impact of new or existing policies on minority health and health disparities may not be evident. More research is needed to analyze local, state, or national health policies that increase or reduce health disparities, such as those related to insurance coverage or reimbursement, organization of government-run or government-funded health-care services, or regulation of environmental hazards. Policies related to zoning or housing, government entitlements, public education, immigration, and criminal justice are needed to be analyzed for effective interventions as well. In order to predict the success or impact of policy initiatives to reduce health disparities, geographic, political, and sociocultural factors should be identified in different regions. Preventive or health improvement interventions should be targeted at nonmedical organizational settings as well, such as schools or the workplace. The feasibility and/or effectiveness of health-care delivery should be evaluated in these nonmedical settings. Since neighborhoods and communities as providers of resources are related to population health and to the production of health inequalities, preventive or health improvement interventions should be evaluated not only at the family but also at the neighborhood or community levels (Bernard et al. 2007). Strategies should be developed to promote greater participation of health disparity populations in health research and clinical trials and the use of health information technologies and/or social media to promote health and health literacy. Patient–provider communication patterns contribute to or minimize health disparities. Improving cultural competence of providers demonstrates a clear link to improvement in patient outcomes, learning patient preferences regarding help-seeking, treatment options, and adherence. More initiatives are needed to increase the supply of health-care practitioners in medically underserved areas.

GIS Tools and Their Role in Targeting Health Inequalities

Health inequalities should be viewed as events spatially empowered with geographic identifiers and signified by a difference in health that deserves scrutiny by applying GIS and appropriate spatial measures. A GIS is a "computer-based system for integrating and analyzing spatially referenced data" (Cromley 2003). GIS is used for mapping disease patterns as well as racial, ethnic, and geographical disparities in health outcomes.

There is compelling evidence that GIS-enabled geography is making a difference in health. We need to know about data and data collection processes and how to integrate health data into GIS. The collection of data on race and ethnicity, for example, is the key step in the implementation of targeted interventions for reducing racial and ethnic disparities in health and heath care. GIS is an indispensable tool in collecting and reporting data on race, ethnicity, and primary language for decision-makers at every level of government and increasingly in the private sector.

GIS is also used for intervention studies, researching the impact of the intervention on existing disparity populations, and for descriptive studies in explaining disparities. If an intervention is to be sensitive to the needs of different subgroups of the population, GIS can be used to know where people with different demographic, ethnic, or socioeconomic characteristics can be found.

GIS tools allow visualization and investigation on many levels of geography such as region, county, zip code, neighborhood, and/or the provider level. GIS enables the use of large geographically referenced data sets, and it encourages research to incorporate comparisons by region and race/ethnicity.

GIS is heavily used for mapping physical environment and developing measures of health outcomes to understand the socioenvironmental determinants of health as indicators for public health policy-making. GIS applications facilitate visual drill down to community and identification of individual hot spots so that we can characterize these populations and identify factors that might contribute to observed inequalities and then target potential interventions that best leverage their available resources. Global positioning system (GPS) and GIS techniques, and the maps they produce, can be valuable assets for in-depth community studies. Maps of the spatial patterns of people's movements enable health-care planners to see at a glance how an intervention strategy would look on the ground (Gesler et al. 2004). For example, GPS is very useful for monitoring physical activity in obesity studies.

There is more potential in encouraging greater use of GIS in clinical trials, using GIS to link large medical data sets. Clinicians can navigate to relevant nonhealth data through real-time GIS. Clinician access to population health data is pertinent to the health status of the communities they serve and how they compare with the larger population so that they can tailor clinical care, outreach, and community services to meet needs better and improve outcomes. GIS, as a means for examining health disparities and findings ways to address them, will take its place in the methodological foundations of population health. My book is organized to foster this understanding.

Organization and Scope

A notable feature of *Spatial Health Inequalities* is the template followed, which consists of chapters, unique to each health disparity topic (i.e., chronic diseases, infectious diseases, etc.), that discusses data and data collection processes, key literature for the appropriate spatial analytical methods applied, and illustrations of cartographic techniques for visualizing and mapping health disparities. It lays out a step-by-step approach for people with an interest in health disparities in their geographic context and draws examples from the United States and the world to identify relevant applications.

This book contains seven chapters. This Introduction explains the term "health inequalities," factors contributing to health inequalities and vulnerable populations affected by inequalities, legacy cases of health inequality, and effectiveness of regulatory interventions. This Introduction also offers a brief overview of GIS and their role in targeting health inequalities. Chapter 1 provides an overview of the current role of GIS in the context of strategic health planning and further reviews the patient-centric and population-oriented uses of GIS. Health inequality topics of chronic diseases, birth health, infectious diseases, and children's lead poisoning are discussed in Chapters 2 through 5. Chapter 6 discusses how GIS is used to evaluate inequalities in health-care access, particularly rural and regional health inequalities and health profession distribution. Chapter 7 is a conclusion chapter that deals with the developments and evolution of GIS in the context of health inequalities.

Each chapter covers a detailed description of a health inequality topic from a global perspective; discusses data collection processes, data types, surveillance techniques, and issues of data quality; and provides key literature for the appropriate spatial analytical methods applied. Illustrations of cartographic techniques for visualizing and mapping that particular health inequality are given. Each chapter concludes with an extensive, updated bibliography.

This book has a web-based resource that provides case studies relevant to each health inequality topic covered in the text. This web page provides links to download data and instructional material to supplement case studies. It aims to be comprehensive in terms of concepts and techniques (but not necessarily exhaustive), representative and independent in terms of modern GIS and related software tools, and above all practical in terms of GIS application and implementation with problem-solving tasks.

The current scholarly literature provides health professionals, preprofessional health students, medical care providers, and medical geographers a broad exposure to public health uses of GIS, emphasizing GIS concepts, methods of disease analysis, and mapping health information, but not unique to each health disparity topic. There is no text helping health professionals to select a body of knowledge to answer specific research questions related to their interest in health inequalities. Adapting GIS to health inequality is a more elusive skill than executing GIS data operational functions. This is a vital book to assist in improving our understanding of the important relationships between people, location, time, and health, therefore assisting in discovering and eliminating health inequalities.

References

Aday, L.A. (2001) *At Risk in America: The Health and Health Care Needs of Vulnerable Populations in the United States*, 2nd edn. San Francisco, CA: Jossey-Bass.

Amick, B.C. (1995). *Society and Health*. New York: Oxford University Press.

Andersen, R. and Newman, J.F. (2005). Societal and individual determinants of medical care utilization in the United States. *Milbank Quarterly* 83(4): 1–28.

Annan, K.A. (2000). *We the Peoples: The Role of the United Nations in the 21st Century*. New York: United Nations Publications.

Bernard, P., Charafeddine, R., Frohlich, K.L., Daniel, M., Kestens, Y., and Potvin, L. (2007). Health inequalities and place: A theoretical conception of neighborhood. *Social Science & Medicine* 65(9): 1839–1852.

Campanelli, R.M. (2003). Addressing racial and ethnic health disparities. *American Journal of Public Health* 93(10): 1624–1626.

CDC. (2013). *Health, United States 2013*. Washington, DC: U.S. Department of Health and Human Services. http://www.cdc.gov/nchs/products/pubs/pubd/hus/state.htm. Accessed November 10, 2015.

Cromley, E.K. (2003). Gis and disease. *Annual Review of Public Health* 24(1): 7–24.

Dall, T., Nikolov, P., and Hogan, P.F. (2004). Economic costs of diabetes in the U.S. in 2002. SSRN Scholarly Paper ID 1154821. Rochester, NY: Social Science Research Network. http://papers.ssrn.com/abstract=1154821. Accessed May 12, 2015.

Evans, T. (2001). *Challenging Inequities in Health from Ethics to Action*. Oxford, England: New York: Oxford University Press.

Frohlich, K.L. and Potvin, L. (2008). Transcending the known in public health practice. *American Journal of Public Health* 98(2): 216–221.

Gesler, W.M., Hayes, M., Arcury, T.A., Skelly, A.H., Nash, S., and Soward, A.C.M. (2004). Use of mapping technology in health intervention research. *Nursing Outlook* 52(3): 142–146.

Grusky, D.B. (1994). *Social Stratification*. Boulder, CO: Westview. http://sites.sdjzu.edu.cn/zhangpeizhong/grusky.pdf. Accessed July 8, 2015.

Heckler, M. (1985). Report of the secretary's task force on black & minority health. http://agris.fao.org/agris-search/search.do?recordID=US201300394700. Accessed February 15, 2015.

Hertzman, C. (1999). The biological embedding of early experience and its effects on health in adulthood. *Annals of the New York Academy of Sciences* 896(1): 85–95.

Iniguez, E. and Palinkas, L.A. (2003). Varieties of health services utilization by underserved Mexican American women. *Journal of Health Care for the Poor and Underserved* 14(1): 52–69.

Institute of Medicine. (2009). *Race, Ethnicity, and Language Data: Standardization for Health Care Quality Improvement*. Washington, DC: National Academies Press.

Kawachi, I., Subramanian, S.V., and Almeida-Filho, N. (2002). A glossary for health inequalities. *Journal of Epidemiology and Community Health* 56(9): 647–52.

Kilbourne, A.M., Switzer, G., Hyman, K., Crowley-Matoka, M., and Fine, M.J. (2006). Advancing health disparities research within the health care system: A conceptual framework. *American Journal of Public Health* 96(12): 2113–2121.

Kuh, D., Shlomo, Y.B. (2004). *A Life Course Approach to Chronic Disease Epidemiology*. New York: Oxford University Press.

Lee, C. (2002). Environmental justice: Building a unified vision of health and the environment. *Environmental Health Perspectives* 110(Suppl 2): 141–144.

Lurie, N. (1997). Studying access to care in managed care environments. *Health Services Research* 32(5): 691–701.

Lurie, N. and Fremont, A. (2006). Looking forward: Cross-cutting issues in the collection and use of racial/ethnic data. *Health Services Research* 41(4 Pt 1): 1519–1533.

Lurie, N., Fremont, A., Somers, S.A., Coltin, K., Gelzer, A., Johnson, R., Rawlins, W., Ting, G., Wong, W., and Zimmerman, D. (2008). The national health plan collaborative to reduce disparities and improve quality. *Joint Commission Journal on Quality and Patient Safety* 34(5): 256–265.

Lurie, N., Jung, M., and Lavizzo-Mourey, R. (2005). Disparities and quality improvement: Federal policy levers. *Health Affairs* 24(2): 354–364.

Lurie, N., Somers, S.A., Fremont, A., Angeles, J., Murphy, E.K., and Hamblin, A. (2008). Challenges to using a business case for addressing health disparities. *Health Affairs* 27(2): 334–338.

Mays, V.M., Cochran, S.D., and Barnes, N.W. (2007). Race, race-based discrimination, and health outcomes among African Americans. *Annual Review of Psychology* 58(1): 201–225.

Murray, C. J. and Lopez. A.D. (1997). Global mortality, disability, and the contribution of risk factors: Global burden of disease study. *Lancet* 349(9063): 1436–1442.

ODPHP—Office of Disease Prevention and Health Promotion. (2000). *Healthy People 2010: Understanding and Improving Health*. Washington, DC: United States Department of Health and Human Services.

Payne-Sturges, D., Gee, G.C., Crowder, K., Hurley, B.J., Lee, C., Morello-Frosch, R., Rosenbaum, A., Schulz, A., Wells, C., Woodruff, T., and Zenick, H. (2006). Workshop summary: Connecting social and environmental factors to measure and track environmental health disparities. *Environmental Research* 102: 146–153.

Perot, R.T. and Youdelman, M. (2001). Racial, *Ethnic, and Primary Language Data Collection in the Health Care System: An Assessment of Federal Policies and Practices*. New York: Commonwealth Fund.

Shi, L. (2000). Vulnerable populations and health insurance. *Medical Care Research and Review* 57(1): 110–134.

Shi, L. (2007). *Health Services Research Methods*, 2d edn. Boston, MA: Delmar Cengage Learning.

Shue, H. (1996). *Basic Rights: Subsistence, Affluence, and U.S. Foreign Policy*. Princeton, NJ: Princeton University Press.

Tufte, E.R. (2001). *The Visual Display of Quantitative Information*, 2nd edn. Cheshire, CT: Graphics Press.

U.S. Congress. (2004). Health Care Equality and Accountability Act. Senate Bill 2091. Section 903a1. http://thomas.loc.gov/cgi-bin/query/z?c109:S.16. Accessed February 12, 2015.

Weeks, W.B., Kazis, L.E., Shen, Y., Cong, Z., Ren, X.S., Miller, D., Lee, A., and Perlin, J.B (2004). Differences in health-related quality of life in rural and urban veterans. *American Journal of Public Health* 94(10): 1762–1767.

Wisner, B. and Adams, J. (2002). *Environmental Health in Emergencies and Disasters: A Practical Guide.* Geneva, Switzerland: World Health Organization.

Zerre, E. and McIntyre, E. (2003). Inequities in under-five child malnutrition in South Africa. *International Journal of Equity in Health*, 2:7.

Zion, D., Gillam, L., and Loff, B. (2000). The Declaration of Helsinki, CIOMS and the ethics of research on vulnerable populations. *Nature Medicine* 6(6): 615–617.

1

New Emerging Trends in Health Information Technology

Given the current emphasis of health-care reform, in the United States and globally, the purpose of this chapter is to provide an overview of the current role of geographic information systems (GIS) in the context of strategic health planning. First, recent changes in the structure of the health-care system and the current role of GIS are described. Second, the patient-centric and population-oriented uses of GIS are reviewed. Last, disease and bioterrorism (BT) surveillance, crisis, and disaster management, as well as strategies for use of interactive maps, are discussed. Conclusions present a research agenda toward a GIS-supported health information technology (HIT) and how we can move beyond the map to view and explore health disparities.

Current Role of GIS in the Health-Care System

There are two big changes taking place in the health-care system in the United States. One of them is controlling the cost of health care by improving the clinical care delivery system through the implementation of an integrated, computer-based, national health-care infrastructure based on an interoperable electronic health record (EHR) system. EHR is the leading record in electronic health-care environment. Documents of the EHR include discharge summaries, radiological images, or lab results, and these documents are available to health-care professionals via special secure web services and/or open-source software. They are built to share information with other health-care providers, such as laboratories and specialists, so they contain information from all the clinicians involved in the patient care (Garrett and Seidman 2014). A full clinical integration (i.e., clinical, acute, and ambulatory care settings) has yet to be achieved, but progress toward integration can be assessed (ONC Data Brief 25 2015).

Second is the recognition of the importance of place in primary care research and practice. Primary care is in the front lines of providing medical care. As healthy lifestyles and chronic disease prevention take a more pivotal role in the U.S. health-care system, the opportunities for integrating the knowledge of individuals and their families' environment into health assessments, decision-making, and treatment have become obvious (Berke 2010). Obesity, diabetes, and heart disease partly originate in busy modern schedules, in the unhealthy food choices available in the stores, and even in the way neighborhoods are designed. Additionally, depression, anxiety, and high blood pressure can arise from chronically stressful conditions at work and home. Doctors, nurses, or other clinicians know that asthma can start in the air around us or from the mold in the walls of our homes. And, just as important, caregivers want to understand how to translate this knowledge into meaningful action. With the efforts to improve outcomes of their patients, primary physicians, for the first time, are really having the chance to broadly explore patient-centered information processing using EHR system.

The Health Information Technology for Economic and Clinical Health (HITECH) Act, a provision of the American Recovery and Reinvestment Act of 2009, was created to stimulate adoption of EHR and support HIT in the United States (Blumenthal 2010). HITECH Act defines HIT as "hardware, software, integrated technologies or related licenses, intellectual property, upgrades, or packaged solutions sold as services that are designed for or support the use by health-care entities or patients for the electronic creation, maintenance, access, or exchange of health information." HITECH Act sets meaningful use of interoperable EHR adoption in the health-care system as a critical national goal (Blumenthal and Tavenner 2010). The anticipated goal is not adoption alone but "meaningful use" of EHRs by providers to achieve significant improvements in care (Kareo 2014). Access to EHR by patients themselves is an explicit expectation in the definition of "meaningful use." Health information exchange (HIE) has emerged as a core capability for hospitals and physicians to achieve "meaningful use" and receive stimulus funding. Health-care vendors are pushing HIE as a way to allow EHR systems to pull disparate data and function on a more interoperable level (Kareo 2014).

The national preoccupation with the cost of clinical and acute care (reducing admissions and adverse events) evident in the lead-up to the daunting task of implementing the national EHR system is well founded, and changes in the health system's pricing, labor, processes, and technology are essential and urgent. However, improving the clinical care delivery system's efficiency and effectiveness will probably have only modest effects on the health of the population. An integration and building of synergy between the best evidence-based interventions at the population level and in the clinical setting are very much needed. Consequently, a GIS with more complete, useful, timely (e.g., real time), and geographically pertinent information can come as a complementary ingredient to HIT. A GIS is a powerful computer mapping and analysis technology that allows large quantities of information to be viewed and analyzed within a geographic context (Miranda et al. 2002). The delineation point between information systems and geographic information systems has been the visualization of data, particularly revealing patterns and trends that are not readily apparent in traditional databases. GIS is far more powerful than its basic mapping abilities, better described as a spatial analytical system that combines computer mapping capability with additional database management and data analysis tools (McLafferty 2003).

Currently, maintaining the integrity and accessibility of EHR on a GIS or GIS-supported HIT is unheard off. GIS has not been incorporated into centralized management of health data. The adoption of EHR is yet to be incorporated in the nationwide health-care system, and the same is true for monitoring and data analysis tools. The development and utilization of effective and real-time efficient EHRs combined with spatial databases can be used for the improvement of the health status and the follow-up of related health parameters. Extending GIS into EHR provides real-time health-care service to patients and allows health-care professionals to explore, identify, and implement preventive measures to inhibit the spread of diseases. The geographic clusters of chronic diseases (i.e., diabetes) could be identified from primary care networks. Public health departments could conduct large-scale real-time public health surveillance. The disease registry data are used for planning individual patient care looking at the different factors within the environment that can affect health-care provision and also disease causes and conducting population-based care. When conducting population-based care, GIS tools expand our understanding of disparities in health outcomes within a community.

Chronic disease management programs and pilot models continue to be tested in the United States to improve the quality of care for patients with chronic conditions. The Chronic Care Model (CCM) serves as the paradigm by which the chronic disease management programs are structured (CDC 2014). The CCM was developed to identify the essential elements of a health-care system that encourage high-quality chronic disease care. These elements include the community, the health system, self-management support, delivery system design, decision support, and clinical information systems. If coupled with GIS capabilities, clinical information systems are likely to become important and valuable contributors to multiple regional data systems targeting improvement in community health. As data communication related to public health continues to develop, more patient-centric clinical settings will become direct contributors to public health databases through electronic data transfer. The success of the information infrastructure at the individual and population levels will enable information flow between different stakeholders in health care to maximize the utility of the information.

Patient-Centric Applications of GIS

Patient-centric GIS approaches focus on the development of information around the patient, in contrast to the approach used by the computerized medical record industry that builds information around each episode or encounter a patient has with the health-care system. The patient-centric uses of GIS are defined as studies highly utilizing patient data from an EHR and/or studies that navigate to relevant nonhealth data through real-time GIS that supports clinical mode of operations, interactive and integrative with other information technologies in health care. The patient-centric uses of GIS highly utilize EHR. EHR provides a rich source of electronically accessible patient data around demographics, race, age, patient address, vital signs, laboratory results, radiology results, and health-care provider clinical documentation. As these detailed data become electronically available, there is an opportunity to aggregate, analyze, and compare its geographic characteristics to related covariables utilizing GIS. The use of GIS in the evaluation of EHR-derived data is of major significance to research as it permits geographic and cofactor-specific targeting of epidemiologic methods and preventive and therapeutic treatment trials for patients with various conditions. This section of the review provides example projects of useful applications of GIS analyses applied to large data sets now available in EHRs. This includes studies that analyze aggregates of data generated during patient visits to clinics/hospitals, diagnostic laboratories, and pharmacies. Individual physician-conducted studies utilizing GIS with data collected at clinical settings are also reviewed. Most studies analyze outcome disparities using GIS in a population of patients.

Projects extending GIS into EHR are rapidly emerging. eHealth-PHINEX is a collaborative project with the University of Wisconsin and Wisconsin Department of Public Health (Guilbert et al. 2011). Their concept focuses on medical record and public HIE. Clinical information in the form of EHR data can be used to inform public health surveillance. In the same way, surveillance can be used to better understand what primary care physicians are seeing in the clinics. They computed economic hardship index using GIS with census indicators (i.e., crowded housing, federal poverty level, unemployment, median income, and percentage of population with less than high school education). Diabetic rates from EHR were compared with public health data and then economic hardship indices were predicted. The researchers concluded that in the Madison area, people with economic hardship are more likely to suffer from chronic diseases.

STARTING POINT is another Wisconsin project looking at a number of strategies that are provided by the Centers for Disease Control and Prevention (CDC) for communities to use to prevent obesity through targeting physical activity and healthier nutrition (University of Wisconsin 2012). The goal of this project is to examine how well the communities are implementing these strategies and then compare those to actual obesity prevalence rates to determine if using those strategies has a significant impact. Obesity data were obtained from a clinical data warehouse. Using census data, obesity prevalence in Dean County was mapped. The study, in agreement with other obesity literature, concluded that higher-income areas had lower obesity rates, while rural communities had higher rates.

New York City has developed a pilot public health program known as NYC Macroscope (Robert Wood Johnson Foundation 2013). The population health surveillance system compiles EHRs from primary care practices to help city health officials monitor and respond to the real-time prevalence of conditions that impact public health. The EHR architecture allows real-time and low-cost data collection.

The Chicago Health Information Technology Regional Extension Center (CHITREC) builds an EHR-enabled community that directly results in collaborative research projects (CHITREC 2014). One of their research interests is geocoding and mapping health data for epidemiology. Chicago Health Atlas is a shared data resource to provide policy makers, researchers, community advocates, and public health leaders insight into the health of the Chicago community and identify opportunities to improve care. They specifically focused on developing tools that balanced the need of privacy for patients and providers, while preserving uniqueness of patients. They obtained Institutional Review Board (IRB) approvals and data extraction is underway at six large health-care institutions throughout Chicago, along with the parallel development of the data visualization platform in GIS. This program extracts diagnoses, medications, and laboratory tests for all patients seen at participating institutions for linking with publicly available citywide data.

In 2007, CDC launched the GIS Surveillance for Heart Disease, Stroke, and Other Chronic Diseases in State and Local Health Departments Project (CDC 2014). The central objective of this training project is to enhance the ability of state health department staff to integrate the use of GIS-informed surveillance into daily operations that support existing priorities for preventing heart disease, stroke, and other chronic diseases. For example, Louisiana's Department of Health and Hospitals enhanced GIS capacity within their agency by participating in this collaborative project provided by CDC. GIS was used to identify parishes with a high burden of cardiovascular disease and the location of Federally Qualified Health Centers that receive federal reimbursement to provide primary and preventive care to medically underserved populations. Louisiana's Chronic Disease Prevention and Control Unit plans to implement the Patient-Centered Medical Home model to deliver quality-driven, cost-effective, culturally appropriate primary care to residents across the state. This model facilitates collaborative partnerships among patients, families, physicians, and other health-care professionals that encourage active participation in care decisions.

Children's National Medical Center has initiated a partnership with Children's national faculty members and the George Washington University GIS team to establish the foundation to improve childhood health in the DC metropolitan region (Jacobs 2014). Patient characteristics from three regional EHRs were obtained: the Children's National Inpatient EHR, the Goldberg Center for Primary Pediatric Care EHR, and the Children's Pediatricians and Associates, which represents 10 independent suburban pediatric practices. Patient-specific data are aggregated and integrated with other geospatial data sets and analyzed with GIS. The first project focused on childhood immunization, the second on thermal burns, and the third one on childhood obesity. The first project examined the relationship between spatial accessibility to pediatric immunization providers and vaccination compliance in low-income, urban population of children. It was concluded that the children with greater spatial accessibility to pediatric vaccination providers were more likely to be up to date with vaccinations. The second study was designed to identify areas in the District of Columbia with an increased number of pediatric burn injuries and to determine demographic and geographic subgroups at risk for these injuries. In recognition of the frequency of burn injuries in Hispanic toddlers living in at-risk neighborhoods, a partnership was created with political and advocacy groups, including the District of Columbia mayor's office on Latino Affairs, to better reach the Spanish-speaking community. The third study provided a unique opportunity to gather accurate and reliable data derived from regional EHRs related to the condition of obesity and compare these data to cultural and environmental factors to explore geospatial relationships. This study identified significant differences between the study populations of inpatient, suburban, and inner city areas, with a greater prevalence of underweight children in the inpatient group and a greater prevalence of overweight, obese, and severely obese children in the inner city population (Jacobs 2014).

A number of projects are taking place dealing with GIS and health, in particular EHR at clinical settings, by physicians. Geraghty et al. (2010) conducted a study to determine whether there was an association between optimal glucose and lipid control with demographic and socioeconomic status (SES) variables (median income, education attainment, unemployment, and white and black race). They used registry data derived from the University of California Davis Health System's electronic medical record system to identify patients with diabetes mellitus from a network of 13 primary clinics in the Greater Sacramento area. Patient variables requested included address, age, sex, primary care provider name and office location, insurance type, race/ethnicity, primary language, and last glycohemoglobin [A1c], last low-density lipoprotein cholesterol, last urine microalbumin/creatinine results. They mapped the diabetic population of the primary care network and the comparison between driving distances of each patient's primary care provider using GIS's mapping capabilities. Since SES was not available as part of patient demographics in their electronic medical records, they also used GIS to obtain SES information from the U.S. Census Bureau about their practice population. Neighborhood SES was associated with A1c levels, with lower-income neighborhoods having higher A1c levels, indicating less controlled diabetes. This was true even though the individual's SES was not associated with A1c levels. This suggested that neighborhoods could affect diabetes control despite the SES of the individual.

Baumgardner et al. (2010) developed a sociocultural model predictive of ADHD diagnosis with his study of eastern Wisconsin children with ADHD diagnosis. The authors indicated that ADHD rates tend to be correlated with male gender, white race, lower block group median household income, and population density and greater distance to the nearest park and school districts, where they aggressively looked for students diagnosed with ADHD. In another Wisconsin study, Reyes et al. (2013) found ADHD diagnosis prevalence varied with school district boundaries in Dane County, Wisconsin, and was more common among blacks, but not predicted by other geographic factors (Reyes et al. 2013). Baumgardner et al. (2006) conducted a GIS-based study in rural Milwaukee on blastomycosis cases, where they found highest amount of disease along low-lying rivers and higher concentrations around waterways and sandy soils.

Odero et al. (2007) described an electronic injury surveillance system that monitored injury incidence and distribution patterns. They integrated geographic coordinates of the injury site into GIS along with patients' clinical data. They created digital maps of injury spatial distribution using GIS software and correlated injury type and location with patients' clinical data. A computerized medical record system, complemented by GIS technology and an injury-specific component, presented a valuable tool for injury surveillance, epidemiology, prevention, and control for communities served by a specific health facility.

Williams et al. (2003) reviewed patient records for burn injuries at the two children's hospitals in St. Louis. They matched patient addresses to census block groups and calculated burn injury rates. They defined an area of high incidence for burn injuries in North St. Louis. Their spatial statistical analysis combined with mapping injury rates provided a detailed level of injury surveillance and allowed for identification of small areas with elevated rates of specific injuries (Williams et al. 2003). The Health-Care Informatics program, at the University of Wisconsin–Milwaukee and Marquette University School of Dentistry, planned to design and develop a new image-based spatial-oriented EMR system for dental surgeries using GIS. Based on the GIS system, a multilayered data model will be developed for dental EMR to integrate the clinical information with image findings (Wu et al. 2006).

Zimeras et al. (2009) developed a real-time efficient EHR environment in which clinical data could be analyzed and combined with statistical analysis and GIS representation introducing disease monitoring. They created pseudo-GIS interactive maps of Samos island in Greece, a presentation of demographic and health information of the patients injured by accidents. The user is given the opportunity to click on a specific area of the map and acquire information regarding various parameters of the accidents according to the selected area. The type of accident may show a possible correlation between the area of the accident and a specific type of accident that appears in high rates in the specific area.

Primary physicians also use GIS to analyze clinic usage patterns and determine utilization. Health-care organizations use GIS to improve their management practices. Physical access to primary health care remains a key issue. Connectivity and distance could be reasons to explain no-shows.

DuMontier et al. (2013) investigated spatial explanations of higher and lower rates of no-shows in their practice. Primary clinicians use the power of GIS to incorporate information on the demand for care as well as the supply of that care. For example, Rindfleisch (2010) investigated insurance coverage, ethnicity, and income, as well as Madison metro bus routes, as environmental barriers for breast cancer screening in her clinic, and then she was also able to provide a map on where to get mammograms throughout the county. To achieve the maximum population-level increase in accessibility to cancer screening, GIS can be used to optimally site a new mammogram facility. Considerable potential exists for GIS to play a key role in rational and cost-effective health service planning and resource allocation. Calculation of distance usage index is a good example of the manner in which GIS can distil information from a microscale clinic usage patterns and use it at a macroscale to extract meaning and facilitate better decision-making (Tanser 2006).

Managing patient care environments within hospitals and medical centers has become an increasingly complex task. Caregivers require critical information that is readily available in a visually streamlined format. Loma Linda University Medical Center in California uses a GIS-based system called the Patient Location and Care Environment System to let caregivers see the physical bed location of each patient and to retrieve demographic and clinical information (ESRI 2014).

Bazemore et al. (2010) integrated clinical data in a primary care network with GIS and generated distribution, service area, and population penetration maps of those clinics. Qualitative evaluation of the responses of primary care clinic leaders, administrators, and community board members was compared to analytic mapping of their clinic and regional population data. Leaders have expressed the opinion that this exercise would be measurably more valuable if all safety net services were similarly mapping their communities and could collaborate on regional planning. However, broader application of these methods is limited by the lack of financial and technical capacity in community health centers and other primary care settings (Bazemore et al. 2010).

EHR also gives the opportunity to support patient-centered care beyond the limits of one health-care institution and even beyond the limits of a nation and/or a state (Brownstein et al. 2008). For example, for cross-border health surveillance, a simple open-source tool aggregates disease indicator numbers from national databases into a common space where incidence and reported deaths of monitored diseases can be seen and compared across sentinel sites and over time (Veegilo 2014). The capacities of countries for cross-border health surveillance vary in (1) the legal and administrative framework; (2) the ability to detect, evaluate, and report risk situations; and (3) the ability to investigate, intervene in, and communicate international health-risk situations. More resources should be directed toward coordinated action among the countries in order to strengthen surveillance and public health monitoring in their border areas (Quirós et al. 2011). In the European Union (EU), the increasing movement of citizens for work, holiday, and study and of patients and health professionals, respectively, seeking or offering health-care calls for a better coordination of health systems and policies across the EU. In response to the challenge, the European Commission has set up a high-level group to strengthen the collaboration between Member States on health services and medical care. This policy brief aims to contribute to the discussion by providing a review of current information and issues relating to cross-border health care in Europe (WHO 2005).

Population-Oriented Uses of GIS

Population-oriented uses of GIS utilize clinical data and self-reported health information and involve far more than public health uses (Boulos et al. 2011a). Public health uses of GIS focus on prevention rather than on treatment. Public health tracks child immunizations, conducts health policy research, establishes service areas, and reports and responds to infectious disease and other health data that require public health emergencies. Public health data sets contain only data that are mandated for survey, reporting, or capture during follow-up.

GIS is heavily used for mapping and developing measures of health outcomes to understand the socioenvironmental determinants of health as indicators for public health policy making. Typical GIS utilization in public health is for auxiliary purposes giving support to data visualization and simple functions including production of thematic maps, basic map overlay operations, geocoding, and buffering (Chung et al. 2004) rather than for more advanced spatial functions such as spatial statistical analyses, spatial smoothing, adjusting disease rates for covariates, adjusting for social and economic deprivation, adjusting for autocorrelation, and spatial clustering (Rushton 2003). End users are not utilizing their full potential; they prefer to use statistical software to do the brunt of the work and GIS for display and reporting purposes.

Alternatively, population-oriented GIS utilizes regional data warehouses (Blewett et al. 2004) that house clinical data from collaborating care providers. Clinical data repositories and regional data warehouses receive useful data from organizations or locations distinct from traditional health-care providers, for example, environmental or veterinary laboratories, nursing homes, or home health-care providers. Clinicians can navigate to relevant non-health data through real-time GIS that supports modes of operation, interactive and integrative with other information technologies in health care. Clinician access to population health data is pertinent to the health status of the communities they serve and how they compare with the larger population so that they can tailor clinical care, outreach, and community services to better meet the needs of their patient population and improve outcomes.

GIS-supported HIT can handle the attributes of the patient-centric and population-related entities and the relationships between them, while ultimately allowing data sharing between clinical care and population health stakeholders.

Population-oriented GIS involve citizen-contributed geographic information, tracking population data and movements using Internet-enabled and location-aware mobile devices such as tablets and smartphones and remote sensors such as home blood glucose monitors (Boulos et al. 2011b). The next section describes types of population-oriented GIS, such as citizen-contributed geographic information, disease and BT surveillance, and crisis/disaster management.

Citizen-Contributed Geographic Information

Crowd-sourced aggregated data sets as a result of self-reported health information lead to useful insights and information about specific populations, diseases, and geographies with regard to health (Boulos et al. 2011a), for example, highlighting clusters of rash symptom in biosurveillance or reporting dead bird cases for targeting mosquito abatement in efforts to suppress the West Nile virus in areas of high bird loss. Social networks are heavily used in population-oriented GIS forecasting illness outbreaks by tracking Facebook and Twitter updates. Sickweather, a sickness forecasting and mapping system, scans social networks for indicators of illness, tracks illnesses, identifies increased occurrences, compares symptoms, determines which viruses are spreading, and recognizes interesting trends related to seasons and geography (Sickweather 2014). GIS-integrated or location-based health-monitoring technology, for example, wearable, at home, in buildings/rooftops, in street furniture, or in mobile vehicle-mounted devices, enables continuous monitoring of patient health status. These devises can report not only exceptional conditions of a patient but also conditions of the environment through weather sensors. These so-called live sensor geo-mash-ups (Boulos et al. 2008) support monitoring, surveillance, and decision-making tasks of various kinds. This could lead to new possibilities of organizing care and treatment more convenient for patients' daily lives and affordable health care for aging societies.

Disease and Bioterrorism Surveillance

With the recent emphasis on BT and worldwide pandemics, there has been increased interest in systematic analysis of disease patterning to determine whether and where there are outbreaks and to provide reaction time for public health agencies. Current global disease surveillance systems are disease specific and vary by the level of commitment to control these diseases. The formal, informal, and *ad hoc* settings of existing disease-specific surveillance systems in diverse geographic regions in the world need an integrated approach so that all surveillance activities are consolidated into a coherent global health information system. GIS plays a key role in providing the infrastructure for integration. Successful disease surveillance requires standardized methodology, prompt data collection, accurate synthesis of the data, continuity over time, and timely dissemination of the resulting information to the health authorities and to the public (Wiafe and Davenhall 2005). GIS provides new advanced analytical and technological tools and a common platform for linking multidisease surveillance databases with spatial and map information. Web-based GIS makes information instantly available across the globe.

Global health surveillance is important for prompt identification of outbreaks and their source, newly emerging and reemerging infectious diseases, hot spots of disease, new cycles of pandemics, and the threats of BT (Castillo-Salgado 2010).

Clinical care system and veterinary data have been shown to be an important source of surveillance information for infectious diseases, small-area health data, and service use patterns to inform public health efforts. GIS can aid in clinical interventions by predicting outcomes before financial commitments are made. GIS can also aid in prioritizing the allocation of resources and improve efficiency.

Preventing disease is the ultimate objective of disease surveillance. Mapping reveals trends, occurrence patterns of diseases, pathogens, hosts, vectors, and intensity of transmission by the vector populations. GIS provides sophisticated data analysis that can be utilized in monitoring activities on humans, vectors, and pathogens and linking this information with environmental factors and control efforts. Many studies use GIS and spatial statistics to help understand the risk factors (associations with environmental exposures and geographic variables) that are associated with disease incidence. Studies of infectious diseases also use GIS and spatial statistics to investigate the influences of climate, land cover, and other environmental variables that affect habitat suitability for the vectors and hosts (Ostfeld et al. 2005). The ability to extend the possibility of using data, primarily used for patient care, to epidemiologic research will have an influence on how the current GIS and spatial epidemiologic studies are applied and how these generate different study designs and methods for data analysis. Correlated with this development, the types of data to be considered continuously will increase and help make disease eradication efforts more effective.

Geographic Characteristics of Bioterrorism Attack

BT, the intentional dispersal of biological agents by terrorists, concerns the international and local intelligence, law enforcement, medical, and public health communities (Cieslak and Eitzen 2000). Efforts in the United States to deal with possible incidents involving these agents in the civilian sector have progressed with September 11 attacks of the twin towers and New Jersey anthrax experience at U.S. mail processing facility in 2011 (Zubieta et al. 2003). The geographic scope of a biological agent can be at local scale (Torok et al. 1997) but also extend beyond country borders at a larger scale and have a global impact. BT attack could involve multiple agents. Two candidate agents are of special concern—smallpox and anthrax. GIS plays an important role to understand the extent of the affected area and the impacted populations after a BT attack. An enterprise GIS (a system allowing multiple user access) could link data from multiple databases at the federal, state, and local levels and help visualize the BT attack and its impact. Space–time component of a BT attack could be analyzed in a GIS. Verification, treatment, and response to a BT attack could also be supported by GIS. The time component in a BT attack is multidimensional (Inglesby et al. 2000). Some examples of the time component in BT are as follows: the time the attack occurred and spread, the time to clean and sustain the infected property and objects, the time the first symptoms associated with a typical BT agent (such as smallpox), the time a quarantine is imposed or relocation is required (Carus 2001), the time social norms break down such as rioting and looting (O'Toole 1999), the time the fear and anxiety transcends to the general public and the duration of the psychological impact of a BT attack, and the ramifications for vulnerable populations such as elderly, children, and pregnant women. BT response efforts could be connected to disease surveillance. Monitoring of these potential threats has been labeled as syndromic surveillance (Henning 2004) and is made easier by the increased use of EHRs and clinical data, which include signs, symptoms, and diagnostic results. GIS could play its part in the simultaneous analysis of a syndromic surveillance data network. A useful syndromic surveillance (Henning 2004) would have to meet the following points: (a) early detection of outbreaks should be employed; (b) follow the size, spread, and tempo of outbreaks and monitor disease trends by a team comprised both medical and community expertise; (c) validation is required, necessitating comparable baseline data; (d) the technology should exist for multiple syndromic surveillance data sets to be immediately analyzed and cross-referenced; and (e) a specific cost should be allocated for BT initiatives rather than other public health programs absorbing it.

Crisis/Disaster Management

GIS is well established in the field of crisis/disaster management for mitigation, response, and recovery interventions (Newsome and Mitrani 1993; Marcello 1995; Radke et al. 2000; Hodgson and Cutter 2001). GIS is used in both macro (city) and micro (neighborhood/individual) decision-making strategies being involved in a disaster response plan (Palm and Hodgson 1992). At the macro level, most cities and states in the United States have emergency operation centers under Homeland Security that are tasked to develop a comprehensive disaster response system. Having these emergency operation centers and national/local mapping agencies that have a regular program of data collection aids disaster recovery and relief efforts. Existing vector data, enhanced by a high-resolution satellite imagery and topographic maps, provide the initial baseline data. Metadata catalog and enterprise GIS system coordinated with mobile GIS data in the field not only provide guidance and accessibility to necessary spatially referenced databases but also establish confidence in the quality of the data for cooperating agencies.

At the micro level, individuals using mobile phone network data through SIM card movements during disasters and outbreaks and Internet search engine analytics offer new approaches that play a major role in crisis/disaster management. The mobile and social web platforms have the potential to generate voluminous crowd-sourced aggregated data sets for crisis/disaster informatics (Boulos et al. 2011a).

Crisis/disaster informatics coupled with GIS plays an important role in guiding emergency responders to affected areas and mapping the impact to coordinate the relief efforts. For example, the organizations make extensive use of GIS in the production of maps, such as maps of injured or displaced people and those for movement of trucks and shipments. Information could be communicated quickly to relief workers scrambling to aid those that are affected. Information on the location and number of survivors, as well as the extent of their injuries, could be communicated in order to provide food, water, and medical supplies. GIS is used in postrecovery and damage assessments and to discover where best to build, or not build, new schools based on population analysis and proximity to health facilities. Although many emergencies are often unpredictable, much can be done to prevent and mitigate their effects as well as to strengthen the response capacity of communities at risk. The less-resourced communities are often those most severely affected by emergencies and the least prepared to manage them. They can benefit from increased investment in emergency preparedness and risk reduction programs. Studies have shown that minorities often suffer disproportionately in the aftermath of a disaster (Bolin 1986; Bolin and Bolton 1986). Neighborhood-level social risks, poor infrastructure of crowded inner city areas, and lack of transport make the poor more vulnerable during response and recovery of a disaster (Cova and Church 1997). Vulnerability is defined by WHO (1999) as "the conditions determined by physical, social, economic and environmental factors or processes, which increase the susceptibility of a community to the impact of hazards." Hill and Cutter (2001) break vulnerability into three risk categories: individual (choices), social (surrounding neighborhood), and biophysical. These risks combine to present a complex vulnerability matrix, comprising social, racial, and economic layers overlaid on the community's physical and service infrastructure. GIS is critical in the identification of vulnerable communities, whether as cohorts or geographic areas, so that appropriate mitigation and response strategies could be formulated. The vulnerability maps allow the agencies to decide on mitigating measures to prevent or reduce loss of life, injury, and environmental consequences before a disaster occurs. An interdisciplinary risk group considers where mitigating measures should be taken before, for example, a flood occurs. Those preparing the maps can overlap flood inundation and slope stability zones with property maps in order to determine which properties and buildings are at risk. They can then notify the landowners and inform them of government subsidies or other support available for undertaking a measure that would protect their homes from potential damage by, for example, water inundation or slope failure. The houses close to the river have to be protected against flooding. If the river floods, the residents will need to be cared for. Once the river subsides, families need financial help to once again make their house a home. Some homes need to be elevated based on flooding zone regulations. All these stages need to be coordinated efficiently by agencies that develop management plans to deal with infrastructure issues and maintain GIS databases for these occurrences.

Interactive Maps and Analysis Tools

Interactive maps on the Internet platform can reveal health inequalities and their connection to environmental disparities and socially disadvantaged areas. The ability to plot the risk of disease, obtain information on where to target programs of disease prevention, and map suspected disease predictors can inform surveillance and intervention programs. The general public can have access to health and environmental information and visualize and analyze that information in new and innovative ways.

Web-enabled GIS or Internet mapping within the realm of both public and population health is an integration of the Internet and GIS in order to cartographically represent disease incidence data in map form. Web-enabled GIS improves access to existing databases. The dynamic link between databases and web-enabled GIS means that updates to data are automatically reflected in maps. Many organizations want to incorporate GIS technology into their standard programs but lack the resources and expertise to build in-house GIS (Leitner et al. 2000). Cloud-based GIS Internet platforms alleviate this problem by providing affordable GIS tools on the Internet in comparison to purchasing and using stand-alone

GIS systems (Wong and Chua 2004). The institutional context affects GIS implementation and public distribution depending upon rules and regulations. As media and platforms change through time, legal and ethical considerations will redefine public access to confidential health information. The decisions about the scale at which data are collected, how data will be aggregated, and what kinds of data will be distributed affect the quality of geographic data available.

Public health organizations such as the National Cancer Institute and CDC have already developed web GIS sites for presentation of information and ensure timely access to their disease surveillance (NCI and CDC 2015). Numerous local governments, community groups, nonprofit organizations, and libraries have also adapted GIS technologies and developed new methods for acquiring, cataloging, and making public information accessible to the communities.

Conclusion

HIT is the basis for more patient-centered and evidence-based medicine with the real-time availability of high-quality information (Zeng et al. 2009). GIS has the potential to be the platform integrating and building synergy between the evidence-based interventions at the population level and in the patient-centric clinical settings.

The shift from the use of health information data from patient care to health-care planning, as well as clinical and epidemiologic research, will transform health statistics and information systems to a more concrete framework for placing measurement in the service of accountability. As HIT is more and more accepted as the foundation for a new model of health-care system, GIS technology will be enhanced to encompass a stronger role in HIT. GIS-supported HIT will be pivotal in measuring health outcomes and understanding nation's health status. This will lead to better assessment of how well the nation's efforts and investments result in improved population health.

Sustained implementation and continued GIS expertise in developing and industrialized countries are also needed. Growing IT teams are increasingly embedding GIS within their standard health-monitoring tools. The need to support and harness GIS for statistical development globally requires the active involvement of health sectors in standardizing their own geospatial information. International health partners strengthen GIS capacity through a coordinated approach, as geography has a shared value to all efforts targeting improved global health and population outcomes (Ebener et al. 2015).

References

Baumgardner, D.J., Knavel, E.M., Steber, D., and Swain, G.R. (2006). Geographic distribution of human blastomycosis cases in Milwaukee, Wisconsin, USA: Association with urban watersheds. *Mycopathologia* 161(5): 275–282.

Baumgardner, D.J., Schreiber, A.L., Havlena, J.A., Bridgewater, F.D., Steber, D.L., and Lemke, A.M. (2010). Geographic analysis of diagnosis of attention deficit/hyperactivity disorder in children: Eastern Wisconsin, USA. *International Journal of Psychiatry in Medicine* 40(4): 363–382.

Bazemore, A., Phillips, R.L., and Miyoshi, T. (2010). Harnessing geographic information systems (GIS) to enable community-oriented primary care. *Journal of the American Board of Family Medicine: JABFM* 23(1): 22–31.

Berke, E.M. (2010). Geographic Information Systems (GIS): Recognizing the importance of place in primary care research and practice. *Journal of the American Board of Family Medicine: JABFM* 23(1): 9–12.

Blewett, L.A., Parente, S.T., Finch, M.D., and Peterson, E. (2004). National Health Data Warehouse: Issues to consider. *Journal of Healthcare Information Management: JHIM* 18(1): 52–58.

Blumenthal, D. (2010). Launching HITECH. *The New England Journal of Medicine* 362(5): 382–385.

Blumenthal, D. and Tavenner, M. (2010). The 'meaningful use' regulation for electronic health records. *The New England Journal of Medicine* 363(6): 501–504.

Bolin, R.C. (1986). Disaster impact and recovery: A comparison of black and white victims. *International Journal of Mass Emergencies and Disasters* 4(1): 35–51.

Bolin, R.C. and Bolton, P. (1986). *Race, Religion, and Ethnicity in Disaster Recovery.* Boulder, CO: University of Colorado.

Boulos, M.N.K., Resch, B., Crowley, D.N., Breslin, J.G., Sohn, G., Burtner, R., Pike, W.A., Jezierski, E., and Chuang, S.K.Y. (2011a). Crowdsourcing, citizen sensing and sensor web technologies for public and environmental health surveillance and crisis management: Trends, OGC standards and application examples. *International Journal of Health Geographics* 10(1): 67.

Boulos, M.N.K., Scotch, M., Cheung, K., and Burden, D. (2008). Web GIS in practice VI: A demo playlist of geo-mashups for public health neogeographers. *International Journal of Health Geographics* 7(July): 38.

Boulos, M.N.K., Wheeler, S., Tavares, C., and Jones, R. (2011b). How smartphones are changing the face of mobile and participatory healthcare: An overview, with example from eCAALYX. *BioMedical Engineering OnLine* 10(April): 24.

Brownstein, J.S., Clark, C.F., Reis, B.Y., and Mandl, K.D. (2008). Surveillance sans frontières: Internet-based emerging infectious disease intelligence and the HealthMap project. *PLoS Medicine* 5(7): e151.

Carus, W.S. (2001). The illicit use of biological agents since 1990. Working paper: Bioterrorism and biocrimes. Washington, DC: Center for Counterproliferation Research, National Defense University.

Castillo-Salgado, C. (2010). Trends and directions of global public health surveillance. *Epidemiologic Reviews* 32(1): 93–109.

Center for Disease Control (CDC). (2014). Building GIS capacity for chronic disease surveillance. http://www.cdc.gov/dhdsp/programs/gis_training.htm. Accessed February 25, 2015.

Chicago Health IT Regional Extension Center (CHITREC). (2014). Research: CHITREC builds an EHR-enabled community that directly results in collaborative research projects. http://chitrec.org/. Accessed February 25, 2014.

Chung, K., Yang, D., and Bell, R. (2004). Health and GIS: Toward spatial statistical analyses. *Journal of Medical Systems* 28(4): 349–360.

Cieslak, T.J. and Eitzen, E.M.J. (2000). Bioterrorism: Agents of concern. *Journal of Public Health Management and Practice* 6: 4.

Cova, T.J. and Church, R.L. (1997). Modeling community evacuation vulnerability using GIS. *International Journal of Geographical Information Science* 11: 763–784.

DuMontier, C., Rindfleisch, K., Pruszynski, J., and Frey, J.J. (2013). A multi-method intervention to reduce no-shows in an urban residency clinic. *Family Medicine* 45(9): 634–641.

Ebener, S., Guerra-Arias, M., Campbell, J., Tatern, A.J., Moran, A.C., Johnson, F.A., Fogstad, H. et al. (2015). The geography of maternal and new born health: The state of the art. *International Journal of Health Geographics* 14: 19.

Environmental Systems Research Institute (ESRI). (2014). A wealth of tools. GIS for health care today and tomorrow. http://www.esri.com/news/arcuser/0499/umbrella.html. Accessed February 26, 2015.

Garrett, P. and Seidman, J. (2014). EMR vs EHR—What is the difference? Health IT buzz. http://www.healthit.gov/buzz-blog/electronic-health-and-medical-records/emr-vs-ehr-difference/. Accessed February 25, 2014.

Geraghty, E.M., Balsbaugh, T., Nuovo, J., and Tandon, S. (2010). Using geographic information systems (GIS) to assess outcome disparities in patients with Type 2 diabetes and hyperlipidemia. *Journal of the American Board of Family Medicine: JABFM* 23(1): 88–96.

Guilbert, T.W., Arndt, B., Temte, J., Adams, A., Buckingham, W., Tandias, A., Tomasallo, C., Anderson, H.A., and Hanrahan, L.P. (2011). The theory and application of UW eHealth-PHINEX, a clinical electronic health record—Public health information exchange. *Wisconsin Medical Journal* 111(13): 124–133.

Henning, K.J. (2004). What is syndromic surveillance? *Morbidity and Mortality Weekly Report* 53: 7–11.

Hill, A. and Cutter, S. (2011). Methods for determining disaster proneness. In S. Cutter (Ed.), *American Hazardscapes—The Regionalization of Hazards and Disasters*. Washington, DC: Joseph Henry Press, pp. 13–36.

Hodgson, M. and Cutter, S.L. (2001). Mapping and the spatial analysis of hazard-scapes. In S.L. Cutter (Ed.), *American Hazardscapes—The Regionalization of Hazards and Disasters*. Washington, DC: Joseph Henry Press, pp. 37–60.

Inglesby, T., Grossman, R., and O'Toole, T. (2000). A plague on your city: Observations from TOPOFF. *Biodefense Quarterly* 2(2): 1–10.

Jacobs, B.R. (2014). Geospatial mapping and analysis of health care conditions in children. Washington, DC: Children's National Medical Center. https://www.himss.org/files/HIMSSorg/content/HIMSS12PhysPosters/BrianJacobs.pdf. Accessed February 24, 2014.

Kareo. (2014). Meaningful Use Resource Center—Kareo. http://www.kareo.com/meaningful-use. Accessed February 25, 2014.

Leitner, H., Sheppard, E., McMaster, S., and McMaster, R. (2000). Modes of GIS provision and their appropriateness for neighborhood organizations—Examples from Minneapolis and St. Paul, Minnesota. *URISA Journal* 12(4): 43–56.

Marcello, B. (1995). FEMA's new GIS reforms emergency response efforts. *GIS World* 8(12): 70–73.

McLafferty, S.L. (2003). GIS and health care. *Annual Review of Public Health* 24: 25–42.

Miranda, M.L., Dolinoy, D.C., and Overstreet, M.A. (2002). Mapping for prevention: GIS models for directing childhood lead poisoning prevention programs. *Environmental Health Perspectives* 110: 949–950.

National Cancer Institute (NCI) and Center for Disease Control (CDC). (2015). The state cancer profiles. http://statecancerprofiles.cancer.gov/about/. Accessed October 2, 2015.

Newsome, D.E. and Mitrani, J.E. (1993). Geographic information system applications in emergency management. *Journal of Contingencies and Crisis Management* 1: 199–202.

Odero, W., Rotich, J., Yiannoutsos, C.T., Ouna, T., and Tierney, W.M. (2007). Innovative approaches to application of information technology in disease surveillance and prevention in Western Kenya. *Journal of Biomedical Informatics* 40(4): 390–397.

ONC Data Brief 25. (2015). Interoperability among U.S. non-federal acute care hospitals. Washington, DC: The Office of the National Coordinator for Health Information Technology.

Ostfeld, R.S., Glass, G.E., and Keesing, F. (2005). Spatial epidemiology: An emerging (or re-emerging) discipline. *Trends in Ecology and Evolution* 20(6): 328–336.

O'Toole, T. (1999). Smallpox: An attack scenario. *Emerging Infectious Diseases* 5: 540–546.

Palm, R.I. and Hodgson, M.E. (1992). Earthquake insurance: Mandated disclosure and homeowner response in California. *Annals of the Association of American Geographers* 82: 207–222.

Quirós, H.M., González, H.R., Vergara, V., and Fernando, J. (2011). An international health proposal to harmonize crossborder health surveillance. *Revista Panamericana de Salud Pública* 30(2): 148–152.

Radke, J., Cova, T., Sheridan, M.F., Troy, A., Mu, L., and Johnson, R. (2000). Application challenges for geographic information science: Implications for research, education, and policy for emergency preparedness and response. *URISA Journal* 12(2): 15–30.

Reyes, N., Baumgardner, D., Simmons, D.H., and Buckingham, W. (2013). The potential for sociocultural factors in the diagnosis of ADHD in children. *Wisconsin Medical Journal* 12(1): 13–17.

Rindfleisch, K.S. (2010). Optimizing access to the medical home: How one clinic turned chronic failure into success. *43rd STFM (Society of Teachers of Family Medicine) Annual Spring Conference*, Vancouver, British Columbia, Canada.

Robert Wood Johnson Foundation. (2013). Electronic health records and the NYC macroscope: Q&A with Carolyn Greene. New Public Health. http://www.rwjf.org/en/blogs/new-public-health/2013/08/electronic_healthre.html. Accessed August 22, 2013.

Rushton, G. (2003). Public health, GIS, and spatial analytic tools. *Annual Review of Public Health* 24: 43–56.

Sickweather. (2014). Recommend remedies for your sick friends. http://www.sickweather.com/. Accessed June 13, 2014.

Tanser, F. (2006). Geographical information systems (GIS) innovations for primary health care in developing countries. *Innovations: Technology, Governance, Globalization* 1(2): 106–122.

Torok, T.J., Tauxe, R.V., Wise, R.P., Livengood, J.R., Sokolow, R., Mauvais, S., Birkness, K.A., Skeels, M.R., Horan, J.M., and Foster, L.R. (1997). A large community outbreak of salmonellosis caused by intentional contamination of restaurant salad bars. *Journal of the American Medical Association* 278: 389–395.

University of Wisconsin. (2012). EMR, GIS and Y-O-U. Department of Family Medicine Statewide Research Forum. Madison, WI: University of Wisconsin School of Medicine and Public Health. http://videos.med.wisc.edu/videos/40424. Accessed October 10, 2012.

Veegilo. (2014). InSTEDD: Innovative support to emerging diseases and disasters. http://instedd.org/technologies/veegilo/. Accessed February 27, 2014.

World Health Organization (WHO). (1999). Community emergency preparedness: A manual for managers and policy-makers. Geneva, Switzerland: World Health Organization.

World Health Organization (WHO). (2005). Cross border health care in Europe. Brussels, Belgium: European Observatory on Health Systems and Politics.

Wiafe, S. and Davenhall, B. (June 2005). Extending Disease Surveillance with GIS. ArcUser. Redlands, CA: ESRI Press.

Williams, K.G., Schootman, M., Quayle, K.S., Struthers, J., and Jaffe, D.M. (2003). Geographic variation of pediatric burn injuries in a metropolitan area. *Academic Emergency Medicine: Official Journal of the Society for Academic Emergency Medicine* 10(7): 743–752.

Wong, S. and Chua, Y. (2004). Data intermediation and beyond: Issues for web-based PPGIS. *Cartographica* 38: 63–79.

Wu, M., Koenig, L., Lynch, J., and Wirtz, T. (2006). Spatially-oriented EMR for dental surgery. *AMIA 2006 Annual Symposium Proceedings*, Washington, DC, p. 1147.

Zeng, X., Reynolds, R., and Sharp, M. (2009). Redefining the roles of health information management professionals in health information technology. *Perspectives in Health Information Management* 6(Summer): 1f. http://www.ncbi.nlm.nih.gov/pmc/articles/PMC2781729/. Accessed October 2, 2015.

Zimeras, S., Diomidous, M., Zikos, D., and Mantas, J. (2009). An electronic health record model for the spatial epidemiological analysis of clinical data. *Materia Socio Medica* 21(2): 103–109.

Zubieta, C.I., Shimmer, R., and Dean, A.G. (2003). Initiating informatics and GIS support for a field investigation of bioterrorism. The New Jersey anthrax experience. *International Journal of Health Geographics* 2: 8.

2

Chronic Diseases

Technological and social changes influence the risk factors to which populations are exposed, shifting the major causes of death and disease from infections to chronic diseases (Koplan 2002). Chronic disease is a long-lasting condition that can be controlled but not cured. Chronic diseases include heart disease, stroke, cancer, chronic respiratory diseases, obesity, and diabetes. Mapping of chronic diseases started with the recognition that environmental factors play an essential role in their etiology. With the assistance of disease maps, low- and high-risk areas can be highlighted, and environmental factors (physical and/or sociocultural) contributing to the process of causation can be related to chronic diseases. Geographic information systems (GIS) analysis can help with monitoring and tracking chronic disease trends; identifying, implementing, and evaluating effective interventions; and providing guidance for chronic disease management programs. This chapter discusses the use of GIS in chronic disease investigations. It aims to expand our understanding of disparities for adverse outcomes associated with chronic diseases in a geographic context. A representative set of chronic diseases such as diabetes, obesity, and cancer are covered. The emphasis is on GIS operations and applications rather than on statistical issues.

Diabetes

Between 1980 and 2013, the world saw a dramatic increase in the prevalence of obesity and diabetes (Shaw et al. 2010; Ng et al. 2013; Imamura et al. 2015), diseases estimated to affect over 13% and 9% of adults worldwide, respectively (Mendis et al. 2015). The number of adults with diagnosed diabetes in the United States nearly quadrupled, from 5.5 to 21.3 million in recent decades (CDC 2015). End-organ complications from diabetes are a substantial source of morbidity and mortality (Geraghty et al. 2010). Therefore, diabetes management is designated as a very important mission by U.S. health systems to improve the health of diabetic populations and minimize the risk of long-term consequences associated with diabetes such as kidney disease, nerve damage, retinal disease, heart disease, and stroke.

Disparities for adverse outcomes associated with diabetes are well documented for racial and ethnic minorities such as poor diabetes control (i.e., poor measures of glycemic control) and high rates of end-organ complications (Peek et al. 2007). The language barrier, poor access to care, low socioeconomic status (LSES), and self-management are some of the factors associated with these disparities. Major differences in diabetes prevalence between African-Americans and whites may simply reflect differences in established risk factors, such as SES, which typically vary by race (Signorello et al. 2007). There have been numerous studies that show diabetes related to ethnicity and race (Harris et al. 1999; Heisler et al. 2003; Kirk et al. 2008; Misra and Lager 2009). Studies also found association between neighborhood SES and diabetes control (Geraghty et al. 2010). High-quality diabetes care starts with effective control of blood glucose, blood pressure, and lipids, specifically understanding blood glucose level ranges ("sugar," measured in mg/dL), A1c levels (i.e., sugar bound to hemoglobin), and LDL cholesterol (i.e., low-density lipoprotein). Geraghty et al. (2010) found that low-income neighborhoods were associated with higher A1c levels, which indicates less controlled diabetes management. Neighborhoods could affect diabetes control despite the SES of the individual (Geraghty et al. 2010).

Linking diabetes registry with GIS helps analyze the outcome disparities in a population of patients with diabetes (Bodenheimer et al. 2002). When it comes to neighborhood variables, GIS is the key to obtain this information about the diabetic population. Since SES is not included as part of patient demographics in diabetes registries, most health systems are factoring SES variables from census into their strategies by putting GIS resources into their disease management. Patient, socioeconomic, and demographic data are joined to a single database using GIS software. Then, studies can seek to determine whether there is an association between optimal glucose and lipid control with demographic and socioeconomic variables obtained from GIS.

There are many opportunities to use GIS analysis with diabetes registries. By mapping the diabetic population of the primary care networks and making comparisons of driving distance to each patient's primary care provider, medical centers could consider whether to expand their health services to the areas of greatest need for their diabetic patients. Regional and temporal trends (i.e., mortality and incidence) and geographic variation in medical and community resources (e.g., diabetes support groups) could be identified through visual analysis of the maps. For example, Figure 2.1 shows diabetes mortality rates by zip code in Shelby County, Tennessee, United States. Maps highlight pockets of high need or regions with a particular target population for diabetes-related interventions (e.g., diabetes education recommended upon diagnosis) (Handelsman 2011). The impact of disease management programs on geographic regions with high diabetes rates and high densities of minority populations could be determined through maps. The programs help individuals from high minority areas self-manage their disease better. This could result in high A1c testing rates in those regions

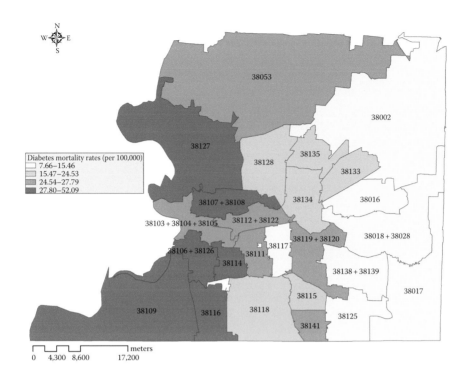

FIGURE 2.1
Diabetes mortality rates (per 100,000) by zip code in Shelby County, Tennessee, 2009–2013. (Courtesy of Memphis and Shelby County Health Department, Memphis, TN.)

(Coberley et al. 2007). Mixed-methods approach using GIS data collected from interviews and surveys helps examine the effectiveness of diabetes programming and resource gaps (Tang et al. 2011). Two survey studies conducted in the state of Michigan found that those living in regions with the highest risk for diabetes were the least likely to receive diabetes screening (Kruger et al. 2008; Curtis et al. 2013). There was one endocrinology practice for every 3774 patients with diabetes in Michigan revealing an uneven distribution, also typical across the United States (Stewart 2008).

GIS has been used in community-based intervention efforts for early diagnosis of diabetes (Geraghty et al. 2010). The U.S. Department of Health and Human Services's (DHHS) Diabetes Detection Initiative (DHHS 2004a) is a community-based effort to identify persons with undiagnosed type 2 diabetes and refer them for follow-up blood testing and treatment. Ten communities throughout the United States have taken the lead in implementing the DHHS's Diabetes Detection Initiative (DHHS 2004b). Zip code areas where diabetes risk rates are high are targeted for Diabetes Detection Initiative activities. These areas are found to have relatively low to moderate diabetes screening rates, where intervention efforts most needed.

There are some challenges of using GIS analysis with diabetes registries. While conducting population-based care and analysis, data reliability is a very important factor. Integrating disease registry data into GIS requires achieving an exact match to patient addresses. A significant number of patients are excluded from the process because of geocoding errors (i.e., post office boxes, address abbreviations, misspellings). Establishing this direct match between registry systems and electronic medical systems early on from the patient's data entry would improve data reliability. While using aggregate data for SES variables, finer scale, less heterogeneous census units, such as census blocks and block groups, may provide a better proxy for the patient's actual SES. Longitudinal data sets provide a better trend analysis. When longitudinal data sets are used, neighborhood change over time also can be monitored. Instead of Euclidian distance, driving time analysis would be more meaningful for analysis. The distinction between Euclidian distance and driving time analysis has significant implications for measuring the size of the problem of poor accessibility to diabetes clinics. The distinction also has implications for the design of policies that may be cost effective in reducing outcome disparities in a population of patients with diabetes.

Obesity

The prevalence of obesity has dramatically risen since the 1980s and has become a global epidemic in the last decade. In 2010, 43 million children were estimated to be obese and overweight; 92 million were at risk of weight gain (Onis et al. 2010). Approximately 16.9% of children in the United States are obese, and minority groups in particular are disproportionately burdened (Ogden et al. 2010; Ogden and Flegal 2010). From 1999 to 2012, 35.1% of African-American children ages 2–19 were overweight, compared with 28.5% of white children (Skinner and Skelton 2014).

Obese children are more likely to become obese adults (Biro and Wien 2010). They are at increased risk of morbidity and mortality (Flegal et al. 2010), disability, premature death due to cardiovascular disease (Freedman et al. 1999), and increased insulin resistance and type 2 diabetes in adulthood (Whitlock et al. 2005).

The development of childhood obesity is multifactorial resulting from a complex set of interactions of these factors. Contributing factors identified include genetics, behavior, environment, and socioeconomic as well as sociodemographic influences. When combined, such factors are perceived to create an environment conducive for obesity development—an obesogenic environment (Mullis et al. 2004; Spiegel and Alving 2005). According to the literature, an obesogenic environment is the totality of influences that the environments have on promoting obesity (Papas et al. 2007; Kirk et al. 2010).

Researchers are beginning to place more emphasis upon the impact that continual environmental exposures have upon obesity development; however, what constitutes an environment that promotes obesity is not well understood. Spatial analysis methods and GIS applications may help understanding of the obesogenic environment phenomenon.

Environmental risk factors associated with obesity are distributed continuously over space, yet the burden of risk may not be evenly dispersed across a community. Geographic disparity must be taken into consideration when examining obesity. For example, the childhood obesity epidemic in the United States appears to disproportionately affect some communities greater than others. A total of 16.5% of rural children qualify as obese as compared with 14.4% of urban children (Simmons 2013). Children living in south-central states of the United States such as West Virginia, Kentucky, Texas, Tennessee, and North Carolina have twice the odds of being obese as compared with children living in the mountain regions such as Colorado, Vermont, and Connecticut (Singh et al. 2008).

Inequality in obesity and its underlying factors, in particular differences in daily physical activity levels and food choices of children living in different environments (i.e., rural, suburban, and urban; high-minority and lower-SES neighborhoods), contribute greatly to health disparity. Neighborhood socioeconomic factors, including poverty and educational level (Drewnowski and Specter 2004), are perceived as fundamental to influencing child weight. Socioeconomically disadvantaged individuals who are less educated and/or make less money have higher obesity rates (Trust for America's Health 2011). Economically, the poorest states in the United States (e.g., West Virginia, Mississippi, and Tennessee) have the heaviest children (Trust for America's Health 2011).

While crime rates tend to be higher in low-socioeconomic communities, studies examining the relationship between neighborhood community crime rates and obesity development have yielded equivocal results (Burdette and Whitaker 2004). Community crime and sex offender rates have been positively associated with indoor sedentary behavior (Brown et al. 2008) as has gender, with girls participating in less physical activity in areas perceived to have higher rates of crime (Gomez et al. 2004). Whether crime, particularly violent crime, contributes to obesity development through greater psychological stress, overeating, or more indoor, sedentary activity remains unclear.

In an effort to better understand the multifaceted nature of childhood obesity, the study of spatial environments may have important implications for health, education, and urban planning policy. Analysis of spatial distribution may help provide answers to fundamental childhood obesity questions related to environmental risks. Features of the built environment are increasingly being recognized as potentially important determinants of obesity. Obesity has been associated with limited community physical resources such as recreational activities, mixed land use, green space, and safety for pedestrians and bicyclists (Cervero and Duncan 2003; Ellaway et al. 2005; Li et al. 2005).

Neighborhoods may have multiple opportunities to access healthy food options but also many competing negative dietary influences such as fast food restaurants and convenience stores. Access to food sources including grocery stores and fruit and vegetable markets appears to reduce the risk of obesity while higher numbers of fast food establishments and convenience stores are linked to obesity development (Blanchard and Lyson 2002; Kipke et al. 2007; Beaulac et al. 2009; Currie et al. 2010). The food insecurity–obesity paradox is also an important factor in health inequalities research. GIS-based studies validate that an abundance of unhealthy food options and food deserts in one's environment leads to lower intake of fruits and vegetables, which can eventually lead to weight gain and obesity. A food desert is a geographic area where affordable and nutritious food is difficult to obtain, particularly for those without access to an automobile (USDA 2015). Some research links food deserts to health inequalities in affected populations (Moore 2003; Li et al. 2005; Powell et al. 2007; Kwate et al. 2009). Figure 2.2 shows an example map of zip codes designated as food deserts.

The ability to measure characteristics of the physical environment and to create a new generation of environmental exposure measures (i.e., the travel

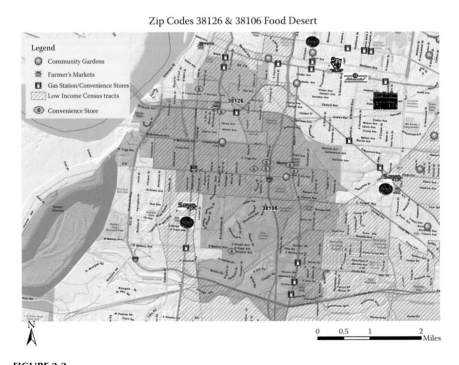

Zip Codes 38126 & 38106 Food Desert

FIGURE 2.2
Food desert areas by zip code in Shelby County, Tennessee, 2010. (Courtesy of City of Memphis Division of Housing and Community Development, Memphis, TN.)

time to the nearest supermarket or calculations of the amount of neighbor-hood green space) is greatly facilitated by the use of GIS (Thornton et al. 2011). GIS methods may provide a means to identify geographical obesity patterns and offer the ability to filter information to identify obesity-related events. Such findings could aid prediction of being overweight and obesity in children and help shape local obesity-prevention interventions. The next section illustrates scale factor, data sources, and obesity measures facilitated by the use of GIS and data limitations.

Scale Factor in Obesity Studies

Reliably identifying the actual size and sociodemographic distribution of the obesity problem represents the first step toward an effective response to childhood obesity. The prevalence of obesity is most often mapped at the coarser scales such as county or state level. While optimal for monitoring international, national, and statewide trends, the WHO's Health Behavior in School-Aged Children survey (WHO 2000; Currie et al. 2004) and current U.S. surveillance systems such as the National Health and Nutrition Examination Survey (NHANES), the Youth Risk Behavior Survey (YRBS 2007), or the National Human Exposure Assessment Survey (Robertson et al. 1998) do not provide detailed objectively assessed data about obesity in "hot spots" such as major urban centers (Kimm and Obarzanek 2002; Obarzanek and Pratt 2002).

As coarser-level data and mapping are not necessarily representative of a community and could potentially conceal patterns of obesity occurrence, there are a variety of finer-scale data sets at the student level, school level, school district level, zip code level (i.e., zip code linked to participant's school or residential address), administrative unit level (e.g., census block groups, census tracts), and buffer level (e.g., circular buffers with varying distances for each participant) that are used in individual/family or school/area analysis, which are part of multilevel data collection, elucidating the potential contribution of various influences on the behavioral pathway to obesity as well as identifying key targets for intervention and policy. Other considerations include scales chosen to represent urban or rural areas, the former representing more localized population and latter utilized by those from a larger spatial region (Macintyre et al. 2002). Buffer sizes generally vary between 100 m and 1 mile (approximately 1600 m) in obesity studies (Robertson-Wilson and Giles-Corti 2010). An advantage of using GIS is that many plausible scales might be easily investigated (Thornton et al. 2011).

At the student level, student demographics are derived from school enrollment data and linked to obesity assessment tests and/or survey material. The assessments usually include obesity and physical activity measurements such as students' height/weight, body mass index (BMI), strength, cardiovascular fitness, endurance, and flexibility. The BMI has been widely utilized to map obesity in epidemiological studies (Borecki et al. 1998).

BMI is computed by weight in kilograms divided by height in meters squared and plotted on a growth chart (Kuczmarski et al. 2002). Percentile cut points developed by the NHANES were utilized for the classification of overweight (BMI ≥ 85th percentile) and obese (BMI ≥ 95th percentile) (Kuczmarski et al. 2000; Barlow and Expert Committee 2007). BMI correlates well with body fat in adults (Bertakis and Azari 2005; Freedman et al. 2005) and children (Dietz and Robinson 1998; Freedman et al. 2006; Velasquez-Mieyer et al. 2007) and is considered a safe, simple, inexpensive method to express relative weight to height to characterize childhood obesity (Himes and Dietz 1994; Freedman et al. 2005).

For mapping purposes, student data had to be aggregated so that they were no longer individually identifiable and therefore could be used to produce maps without extensive legal protection. To guide school- and community-based interventions, specific child obesity projects generate density and proportion maps that graphically display BMI, fitness, and racial/ ethnic and socioeconomic information at the school districts, zip code levels, or buffers derived from school locations. Maps and geographic analyses can provide information on whether obesity rates coincide with levels of economic disadvantage and prevalence of minority students such as Hispanic and African-American students. Residential density maps show geographic areas where concentrations of overweight or obese student populations reside. Residential proportion maps show "hot spots" with high ratios of particular subpopulations (e.g., overweight or obese student populations) to a full population of all students in a particular school or school system.

In addition to school data and school-level analysis, hospitals and healthcare systems have been the site of numerous epidemiological and clinical studies of obesity in both adults (Eck et al. 1992; Alfano et al. 2002; Shen et al. 2006) and children such as school-age children, minority adolescents, and special-needs and at-risk children (Beech et al. 2004; Cooper et al. 2006). They often conduct clinical trials, extensive formative assessments including informant interviews and focus groups with children and parents, and feasibility studies that test various components of their intervention activities. In order to explore the relationship of obesity and being overweight to environmental attributes, these studies also use GIS functions. Such functions include geocoding respondent location buffer analyses drawing buffers around each respondent location and capturing relevant physical activity facilities and eating places (i.e., healthy/unhealthy food choices), computing proximity (e.g., Euclidean and/or network distance to recreation and physical activity places), measuring accessibility (e.g., geometric and population-weighted centroids for potential populations exposed), measuring connectivity and walkability (e.g., amount of intersections, more route choices) and computing density measures (e.g., counts of features such as fast food restaurants), and aggregating to census units to be able to link respondents to primary indicators of SES (education level, ethnic minority populations, income), particularly associated with obesity and physical activity).

Data Resources

In order to explore the relationship of obesity and being overweight to environmental attributes (e.g., socioeconomic, physical, and personal environmental attributes), studies rely on primary and secondary data sources. One of the greatest challenges facing GIS users is the acquisition of detailed data sources that contain locational and attribute information on the built environment (Thontorn et al. 2011). Primary data are based on surveys of individuals who report on characteristics of environmental features (Troped et al. 2001; Mujahid et al. 2007) or field observations undertaken by fieldwork auditors who visit neighborhoods (Raudenbush and Sampson 1999; Pikora et al. 2002), and data from Global Positioning Systems (Stopher et al. 2008) and remote sensing using, for instance, satellites or aerial photographs to identify green space, topography, and so on.

Secondary spatial data come from diverse sources such as administrative data (e.g., from a census), crime data (e.g., from local police departments), emergency food sources and hunger-related data (e.g., from Food Bank), commercially purchased data (e.g., business data from market research companies), transportation networks and infrastructure (e.g., from a state Department of Transportation), and street network data (e.g., from TIGER/Line files from the U.S. Census Bureau). Compared to primary data, secondary data sources are relatively cost effective to obtain and can usually be sourced for specific study areas or across a large geographical area (e.g., the North American Industry Classification System [NAICS]).

Given the breadth of measures of environmental attributes used across obesity studies, secondary data are synthesized under three domains: socioeconomic, physical, and personal attributes. Examples of data sources for each domain are provided as follows.

Socioeconomic Data

U.S. Census and the American Community Survey (ACS) demographics are two important data sources highly used for obesity measures at the neighborhood level (i.e., census units, sociodemographic, and housing data). Household income, education, and minority population data could be obtained from the U.S. Census Bureau (1-year, 3-year, and 5-year) estimates derived from the ACS. The National School Lunch Program is another measure of LSES, coded by school and aggregated to upper levels of geography (e.g., county).

Physical Environmental Data

Business data facilitate neighborhood asset mapping by locating health-care facilities, sources of healthy or unhealthy food (e.g., fast food restaurants), and fitness and recreational resources (gyms, swimming centers, parks, play grounds, etc.). Business data sets (e.g., Info USA) are usually a commercially purchased set of standards. The NAICS includes data summarized into a single measure of all facilities and subdivided into smaller categories of specific types of facilities (e.g., eating places, outdoor facilities).

Police incident data obtained from local police departments are used to map crimes that are likely to diminish personal safety and restrict outdoor activity, whether for recreation or while walking to school.

Transport infrastructure data such as the presence of sidewalks and controlled intersections, road hazards, crossings, access to destinations, and public transportation as well as number of roads to cross, traffic signage, and traffic density/speed are used in studies investigating the relationship between the built environment and physical activity of children (Braza et al. 2004; Timperio et al. 2004; Boarnet et al. 2005). Local planning departments, the Metropolitan Planning Organizations, and state departments of transportation and safety in the United States collect these data and provide in a GIS format for mapping and spatial analysis purposes. In an evaluation of the implementation of a Safe Routes to School program in the United States, children's rates of walking and cycling to school are identified, and that variable is used to find associations with provision of amenities such as the presence and condition of sidewalks and bike lanes (Boarnet et al. 2005; Jago et al. 2005; Mota et al. 2005).

Personal Environmental Attributes

Personal environmental variables include self-reports of the perceived environment (children's and adults', usually parents) in addition to family history (e.g., diabetes, heart disease), and child fitness characteristics. Family history and child fitness data (e.g., percentage of total fifth grade class achieving healthy fitness) are derived from questionnaires and mapped at the zip code, school, and/or county level. Data related to people's perception play an important role in obesity studies incorporating GIS with their mixed-methods approach (Mota et al. 2005).

Data Limitations

As the use of GIS becomes more common place, metadata (information about each data set) documentation is imperative in obesity studies. The various processing decisions when using GIS should be included in published research. The inclusion of such information helps compare findings across studies. GIS has the potential to facilitate the establishment of a transdisciplinary approach and contribute to the advancement of our understanding of the importance of the built environment for obesity. However, important methodological challenges remain relating to the measurement of the built environment (Matthews et al. 2009; Oakes et al. 2009; Saelens and Glanz 2009; Story et al. 2009; Amarasinghe and D'Souza 2010).

GIS is a digital representation of the environment where attributes are objectively measured but GIS cannot take place of perceived environment, for example, children's behaviors from their trip to school. Mixed-methods approach incorporating GIS analysis and people's perceptions should also be taken into consideration in obesity research.

Most data brought into GIS for obesity research are obtained from secondary data resources and not designed for the analytical purposes for which they are being used; therefore, researchers have limited control over much of the data collection processes for their studies (Mainous and Hueston 1997; Nicoll and Beyea 1999; Burns and Groves 2001). In order to ensure secondary data accuracy, validation against primary data is often preferable. Issues related to the accuracy of secondary data sources have previously been reported for physical activity facilities (Boone et al. 2008; Paquet et al. 2008, Giles-Corti et al. 2005) and the food environment (Lake and Townshend 2006; Paquet et al. 2008). Examples of this are that facilities included in commercial databases were not found in the field, or facilities found in the field were not in the commercial database (Boone et al. 2008). Another example of this is that some facilities identified in the field were not considered to be the same service type as that listed in the database (Boone et al. 2008).

Contribution of spatial autocorrelation and transboundary spillover due to activities in one community influencing activities in other communities (Amarasinghe et al. 2006) should also be taken into consideration while studying geographic variation of obesity. Zip codes are administrative entities of the U.S. Postal Service. As such, boundaries are implied and may not necessarily follow clearly identifiable physical or social features.

Cancer

GIS's Role in Cancer Health Disparities

Cancer researchers currently use GIS in ecologic surveillance (i.e., monitoring patterns of rates reflecting interventions); identification of cancer health disparities, geographically focusing cancer control efforts (e.g., disseminating state- or county-level cancer information); evaluation of cancer control interventions; predictive modeling of cancer rates; identification of cancer clusters; planning of health-care resources; and determining ecological correlations (i.e., impact of cancer control interventions). The National Cancer Institute (NCI) in the United States has a long-standing interest in the geographic patterns of cancer, particularly for communication of georeferenced cancer statistics (Mason 1975). There have been successes in publishing cancer maps and development of web-based cancer mapping tools (i.e., gis.cancer.gov). The NCI has a growing program in GIS, actively engaged in projects in the areas of spatial data analysis, and GIS database and geovisualization tools development.

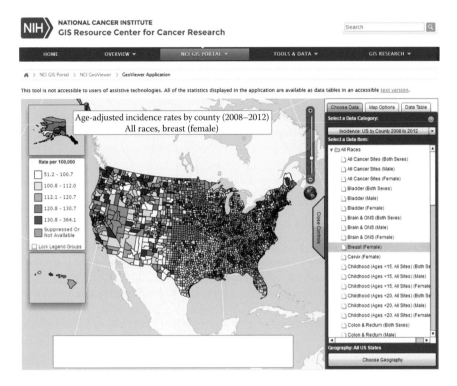

FIGURE 2.3
A screen shot from the National Cancer Institute GeoViewer Application. The map on the left has been created from the State Cancer Profiles data on the right. (From CDC's National Program of Cancer Registries Cancer Surveillance System (NPCR-CSS), Rates are age-adjusted to the 2000 US standard population using SEER*Stat, Data submission: January 2015 and SEER submission: November 2014. http://seer.cancer.gov.; https://gis.cancer.gov/geoviewer/data.)

Online, interactive health atlases, such as the State Cancer Profiles website (NCI and CDC 2015) are designed to facilitate data querying and exploration for a wider audience and communicate geographic variation. For example, Figure 2.3 is a resultant map of a query of female breast cancer on the State Cancer Profiles through NCI Geoviewer application.

Addressing cancer health disparities, such as higher cancer death rates, less frequent use of proven screening tests, and higher rates of advanced cancer diagnoses in certain populations (Klassen et al. 2015), is an area in which the use of GIS has not been ventured into very far as of yet. These disparities are frequently seen in people from low-socioeconomic groups, certain racial/ethnic populations, and from those who live in geographically isolated areas (Ward et al. 2004). For example, in the United States, African-American race (Miller et al. 2002; Newman 2005; Morris et al. 2008; Cunningham et al. 2013) and low-SES population (Simon and Severson 1997; Thomson et al. 2001) are consistently associated with more adverse disease characteristics, including advanced stage at diagnosis, larger tumor size (Miller et al. 2002), and more aggressive tumor biology (Simon and Severson 1997; Cunningham et al. 2012). A recent report on the status of cancer in the United States showed large variations of cancer incidence and death by race and ethnicity (Kohler et al. 2015).

The modifiable cancer risk factors that vary by race/ethnicity and SES include cigarette smoking, physical inactivity, and obesity. Poor and minority communities are selectively targeted by the marketing strategies of tobacco companies (Haiman et al. 2006), may have limited access to fresh foods and healthy nutrition (Freedman and Bell 2009; Larson et al. 2009), and are provided with fewer opportunities for safe recreational physical activity (Davison and Lawson 2006). Income, education, and health insurance coverage influence access to appropriate early detection, treatment, and palliative (i.e., soothing) care. Social inequities such as the legacy of racial discrimination in the United States can still influence the interactions between patients and physicians (Nelson et al. 2002).

Geographic differences in the types or burden of cancer could be explained by regional variations in the prevalence of major risk factors, availability and use of medical practices such as cancer screening, availability and quality of treatment, cultural practices, and age structure (Garcia et al. 2007). For example, resources for cancer therapy are limited in low- and lower-middle-income countries. Surgeons skilled in cancer surgery are in short supply and more likely to be available in urban rather than rural regions. Radiation therapy has still not spread to all countries in the world, in part because of the capital costs of equipment, but also because of the lack of radiation oncologists and medical physicists (Garcia et al. 2007). Cultural factors also play a role in health behaviors, attitudes toward illness, and faith in modern medicine versus alternative forms of healing (Freeman 2003).

As cancer is an effect due to toxic mechanisms operating in nonreproductive cells, understanding the links between environmental pollutants (e.g., carcinogens) and cancer is critical for cancer health disparities research. GIS and spatial statistical techniques used to model exposure and measure environmental justice determinants allow health researchers to describe and understand more fully the nature of communities with excess cancer burden. In the United States, racial and ethnic minorities disproportionately live near toxic waste sites (Chakraborty 2011). They are also more likely to live in areas of high industrialization, air and water pollution, and work in environments that expose them to cancer-causing toxicants. Moreover, this same group of people has a higher rate of exposure and usage of insecticides and pesticides through agriculture work or use in the home. Typically, environmental justice studies do not attempt to explicitly define the spatial distribution of environmental risk as it is an extremely complex task that differs according to type of facility, type of toxic release, and a host of environmental variables that control the dispersion of the toxic material through the environment. Instead, most studies simply consider the people in the "community" (whether defined by census-based areal unit or distance buffer) that hosts the hazardous facility to be the cause of the increased risk (Chakraborty 2011; Chakraborty et al. 2012).

Given the growing global burden of cancer, differing significantly by geography, ethnicity, and socioeconomic status, primary prevention from both behavioral and environmental perspectives is emerging as a critical strategy. In addition to prevention efforts, researchers seek to reduce the burden of cancer by studying interventions, their impact in defined populations, and the means by which they can be better used (Pickle et al. 2006). Identifying where the cancer burden is elevated leads to discovery of locations where interventions are needed. Figure 2.4 is an example of a choropleth map commonly used in community discussions about local cancer rates and cancer control planning.

GIS mapping of cancer patterns may reveal synergistic effects of multiple social, behavioral, environmental, and cultural influences that may not come to light in studies of these same variables outside the geographic context (Howe 1989). Linking population-based surveillance data by GIS to attributes of the neighborhood environment, both individual and community influences on cancer patterns could be investigated within specific clinical populations. When individual-level data are not available, GIS analyses can identify

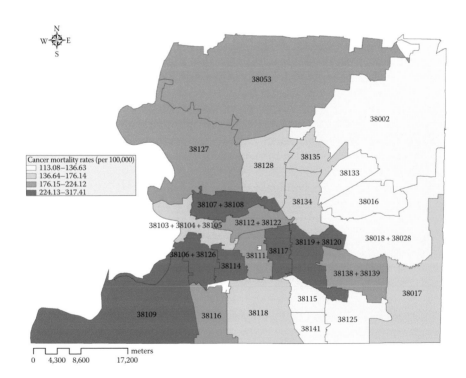

FIGURE 2.4
Cancer mortality rates by zip code in Shelby County, Tennessee, 2009–2013. (Courtesy of Memphis and Shelby County Health Department, Memphis, TN.)

contextual characteristics of populations (i.e., group behaviors and norms, crime, green space, food deserts, pollution, etc.) that influence all members of communities (Loomis et al. 2013). However, GIS-based ecological studies are limited to cross-sectional associations at time of diagnosis, giving no indication of the sequence of events—whether exposure occurred before, after, or during the onset of the disease outcome. This being so, it is impossible to infer causality, though with high sample sizes, this problem can possibly be overcome.

Conclusion

As many of these studies point out, the important issue is how our improved understanding of the chronic disease process leads to better prevention and intervention and improves access to health services. GIS plays an important role in intervention efforts and in understanding evolving patterns of accessibility and their consequences.

There is a growing body of research suggesting that some chronic conditions may be partly the result of infectious agents (Ewald and Cochran 1999). Since infectious diseases are often undiagnosed, untreated, and unreported, delays in diagnosis and treatment often result in severe chronic health problems. The interaction between human, environment, and infectious agents is another level of information that needs to be considered to understand chronic diseases. The role of GIS in analyzing spatial epidemiology of infectious diseases is considered in Chapter 4.

References

Alfano, C.M., Klesges, R.C., Murray, D.M., Beech, B.M., and McClanahan, B.S. (2002). History of sport participation in relation to obesity and related health behaviors in women. *Preventive Medicine* 34(1): 82–89.

Amarasinghe, A., D'Souza, G., Brown, C., and Borisova, T. (2006). The impact of socioeconomic and spatial differences on obesity in West Virginia. In *Annual Meeting*, Long Beach, CA, Vol. 23, p. 26.

Amarasinghe, A. and D'Souza, G.E. (2010). *Economics of Health, Obesity and the Built Environment: Empirical Investigations and Policy Implications.* Saarbrucken, Germany: VDM Verlag.

Barlow, S.E. (2007). Expert committee and treatment of child and adolescent overweight and obesity: Expert committee recommendations regarding the prevention. *Pediatrics* 120(Suppl. 4): S164–S192.

Beaulac, J., Kristjansson, E., and Cummins, S. (2009). A systematic review of food deserts, 1966–2007. *Preventing Chronic Disease* 6(3): A105.

Beech, B.M., Kumanyika, S.K., Baranowski, T., Davis, M., and Roninson, T.N. (2004). Parental cultural perspectives in relation to weight-related behaviors and concerns of African-American girls. *Obesity Research* 12(Suppl.): S7–S19.

Bertakis, K.D. and Azari, R. (2005). The impact of obesity on primary care visits. *Obesity Research* 13(9): 1615–1623.

Biro, F.M. and Wien, M. (2010). Childhood obesity and adult morbidities. *American Society for Nutrition* 91(5): 1499s–1505s.

Blanchard, T. and Lyson, T. (2002). Access to low cost groceries in nonmetropolitan counties: Large retailers and the creation of food deserts. *Measuring Rural Diversity Conference Proceedings*, Washington, DC, November 2002, pp. 21–22.

Boarnet, M.G., Anderson, C.L., Day, K., McMillan, T., and Alfonzo, M. (2005). Evaluation of the California Safe Routes to School Legislation: Urban form changes and children's active transportation to school. *American Journal of Preventive Medicine* 28(2 Suppl. 2): 134–140.

Bodenheimer, T., Wagner, E.H., and Grumbach, K. (2002). Improving primary care for patients with chronic illness. *JAMA* 288: 1775–1779.

Boone, J.E., Gordon-Larsen, P., Stewart, J.D., and Popkin, B.M. (2008). Validation of a GIS facilities database: Quantification and implications of error. *Annals of Epidemiology* 18: 371–377.

Borecki, I.B., Higgins, M., Schreiner, P.J., Arnett, D.K., Mayer-Davis, E., and Hunt, S.C. (1998). Evidence for multiple determinants of the body mass index: The national heart, lung, and blood institute family heart study. *Obesity Research* 6(2): 107–114.

Braza, M., Shoemaker, W., and Seeley, A. (2004). Neighborhood design and rates of walking and biking to elementary school in 34 California communities. *American Journal of Health Promotion* 19(2): 128–136.

Brown, H.S., Pérez, A., Mirchandani, G.G., Hoelscher, D.M., and Kelder, S.H. (2008). Crime rates and sedentary behavior among 4th grade Texas school children. *International Journal of Behavioral Nutrition and Physical Activity* 5(1): 28.

Burdette, H.L. and Whitaker, R.C. (2004). Neighborhood playgrounds, fast food restaurants, and crime: Relationships to overweight in low-income preschool children. *Preventive Medicine* 38(1): 57–63.

Burns, N.B. and Groves, S.K. (2001). *The Practice of Nursing Research: Contact, Critique and Utilization*, 4th edn. Philadelphia, PA: W.B. Saunders.

CDC's National Program of Cancer Registries Cancer Surveillance System (NPCR-CSS). Rates are age-adjusted to the 2000 US standard population using SEER*Stat. Data submission: January 2015 and SEER submission: November 2014. http://seer.cancer.gov; https://gis.cancer.gov/geoviewer/data.

Centers for Disease Control and Prevention (CDC). (2015). Diabetes report card 2014. Atlanta, GA: Centers for Disease Control and Prevention, U.S. Department of Health and Human Services.

Cervero, R. and Duncan, M. (2003). Walking, bicycling, and urban landscapes: Evidence from the San Francisco Bay area. *American Journal of Public Health* 93(9): 1478–1483.

Chakraborty, J. (2011). Cancer risk from exposure to hazardous air pollutants: Spatial and social inequities in Tampa Bay, Florida. *International Journal of Environmental Health Research* 22(2): 165–183.

Chakraborty, J., Maantay, J.A., and Brender, J.D. (2012). Disproportionate proximity to environmental health hazards: Methods, models, and measurement. *American Journal of Public Health* 101(S1): S27–S36.

Coberley, C.R., Puckrein, G.A., Dobbs, A.C., McGinnis, M.A., Coberley, S.S., and Shurney, D.W. (2007). Effectiveness of disease management programs on improving diabetes care for individuals in health-disparate areas. *Disease Management*, 10(3): 147–155.

Cooper, T.V., Klesges, L.M., DeBon, M., Klesges, R.C., and Shelton, M.L. (2006). An assessment of obese and non obese girls' metabolic rate during television viewing, reading, and resting. *Eating Behaviors* 7(2): 105–114.

Cunningham, J.E., Walters, C.A., Hill, E.G., Ford, M.E., Barker-Elamin, T., and Bennett, C.L. (2012). Mind the gap: Racial differences in breast cancer incidence and biologic phenotype, but not stage, among low-income women participating in a government-funded screening program. *Breast Cancer Research and Treatment* 137(2): 589–598.

Currie, C., Roberts, C., Morgan, A., Smith, R., Setterbulte, W., Samdal, O., and Rasmussen, V.B. (Eds.). (2004). Young people's health in context. Health behavior in school-aged children: A WHO Cross National Collaborative Study (HBSC International Report from the 2001/2 Survey). Health Policy for Children and Adolescent, No.4. Geneva, Switzerland: World Health Organization.

Currie, J., Vigna, S.D., Moretti, E., and Pathania, V. (2010). The effect of fast food restaurants on obesity and weight gain. *American Economic Journal: Economic Policy* 2(3): 32–63.

Curtis, A.B., Kothari, C., Paul, R., and Connors, E. (2013). Using GIS and secondary data to target diabetes-related public health efforts. *Public Health Reports* 128(3): 212–220.

Davison, K.K. and Lawson, C.T. (2006). Do attributes in the physical environment influence children's physical activity? A review of the literature. *International Journal of Behavioral Nutrition and Physical Activity* 3(1): 19.

DHHS. (2004a). Diabetes detection initiative. http://www.ndep.nih.gov/ddi. Accessed November 10, 2015.

DHHS. (2004b). Diabetes detection initiative: Finding the undiagnosed. Washington, DC: U.S. Department of Health and Human Services. http://www.ndep.nih.gov/ddi/about/index.htm. Accessed October 8, 2015.

Dietz, W.H. and Robinson, T.N. (1998). Use of the body mass index (BMI) as a measure of overweight in children and adolescents. *Journal of Pediatrics* 132: 191–193.

Drewnowski, A. and Specter, S.E. (2004). Poverty and obesity: The role of energy density and energy costs. *The American Journal of Clinical Nutrition* 79(1): 6–16.

Eck, L.H., Klesges, R.C., Hanson, C.L., and Slawson, D. (1992). Children at familial risk for obesity: An examination of dietary intake, physical activity and weight status. *International Journal of Obesity and Related Metabolic Disorders: Journal of the International Association for the Study of Obesity* 16(2): 71–78.

Ellaway, A., Macintyre, S., and Bonnefoy, X. (2005). Graffiti, greenery, and obesity in adults: Secondary analysis of European cross sectional survey. *British Medical Journal* 331(7517): 611–612.

Ewald, P.W. and Cochran, G. (1999). Catching on to what's catching. *Natural History* 108(1): 34–37.

Flegal, K.M., Carroll, M.D., Ogden, C.L., and Curtin, L.R. (2010). Prevalence and trends in obesity among US adults, 1999–2008. *Journal of the American Medical Association* 303(3): 235–241.

Freedman, D.A. and Bell, B.A. (2009). Access to healthful foods among an urban food insecure population: Perceptions versus reality. *Journal of Urban Health* 86(6): 825–838.

Freedman, D.S., Dietz, W.H., Srinivasan, S.R., and Berenson, G.S. (1999). The relation of overweight to cardiovascular risk factors among children and adolescents: The Bogalusa Heart Study. *Pediatrics* 103: 1175–1182.

Freedman, D.S., Khan, L.K., Serdula, M.K., Dietz, W.H., Srinivasan, S.R., and Berenson, G.S. (2005). The relation of childhood BMI to adult adiposity: The Bogalusa Heart Study. *Pediatrics* 115: 22–27.

Freedman, D.S., Khan, L.K., Serdula, M.K., Ogden, C.L., and Dietz, W.H. (2006). Racial and ethnic differences in secular trends for childhood BMI, weight, and height. *Obesity* 14(2): 301–308.

Freeman, H.P. (2003). Commentary on the meaning of race in science and society. *Cancer Epidemiology Biomarkers & Prevention* 12(3): 232s–236s.

Garcia, M., Jemal, A., Ward, E.M., Center, M.M., Hao, Y., Siegel, R.L., and Thun, M.J. (2007). *Global Cancer Facts & Figures 2007*. Atlanta, GA: American Cancer Society, Vol. 1(3), 52pp.

Geraghty, E.M., Balsbaugh, T., Nuovo, J., and Tandon, S. (2010). Using Geographic Information Systems (GIS) to assess outcome disparities in patients with type 2 diabetes and hyperlipidemia. *Journal of the American Board of Family Medicine* 23(1): 88–96.

Giles-Corti, B., Timperio, A., Bull, F., and Pikora, T. (2005). Understanding physical activity environmental correlates: Increased specificity for ecological models. *Exercise and Sport Sciences Reviews* 33(4): 175–181.

Gomez, J.E., Johnson, B.A., Selva, M., and Sallis, J.F. (2004). Violent crime and outdoor physical activity among inner-city youth. *Preventive Medicine* 39(5): 876–881.

Haiman, C.A., Stram, D.O., Wilkens, L.R., Pike, M.C., Kolonel, L.N., Henderson, B.E., and Le Marchand, L. (2006). Ethnic and racial differences in the smoking-related risk of lung cancer. *New England Journal of Medicine* 354(4): 333–342.

Harris, M., Eastman, R.C., Cowie, C.C., Flegal, K.M., and Eberhardt, M.S. (1999). Racial and ethnic differences in glycemic control of adults with type 2 diabetes. *Diabetes Care* 22: 403–408.

Heisler, M., Smith, D.M., Hayward, R.A., Krein, D.L., and Kerr, E.A. (2003). Racial disparities in diabetes care processes, outcomes, and treatment intensity. *Medical Care* 41: 1221–1232.

Himes, J.H. and Dietz, W.H. (1994). Guidelines for overweight in adolescent preventive services: Recommendations from an expert committee. *American Journal of Clinical Nutrition* 59(2): 307–316.

Howe, G.M. (1989). Historical evolution of disease mapping in general and specifically of cancer mapping. *Recent Results in Cancer Research* 114: 1–21.

Imamura, F., Micha, R., Khatibzadeh, S., Fahimi, S., Shi, P., Powles, J., and Global Burden of Diseases Nutrition and Chronic Diseases Expert Group (NutriCoDE). (2015). Dietary quality among men and women in 187 countries in 1990 and 2010: A systematic assessment. *The Lancet Global Health* 3(3): e132–e142.

Jago, R., Branowski, T., Zakeri, I., and Harris, M. (2005). Observed environmental features and the physical activity of adolescent males. *American Journal of Preventive Medicine* 29(2): 98–104.

Kirk, J.K., Passmore, L.V., Bell, R.A. et al. (2008). Disparities in A1c levels between Hispanic and non-Hispanic white adults with diabetes: A meta-analysis. *Diabetes Care* 31: 2400–2406.

Kirk, S.F., Penney, T.L., and McHugh, T.L. (2010). Characterizing the obesogenic environment: The state of the evidence with directions for future research. *Obesity Review* 11(12): 109–117.

Kimm, S.Y. and Obarzanek, E. (2002). Childhood obesity: A new pandemic of the new millennium. *Pediatrics* 110(5): 1003–1007.

Kipke, M.D., Iverson, E., Moore, D., Booker, C., Ruelas, V., Peters, A.L., and Kaufman, F. (2007). Food and park environments: Neighborhood-level risks for childhood obesity in east Los Angeles. *Journal of Adolescent Health* 40(4): 325–333.

Klassen, A.C., Pankiewicz, A., Hsieh, S., Ward, A., and Curriero, F.C. (2015). The association of area-level social class and tobacco use with adverse breast cancer characteristics among white and black women: Evidence from Maryland, 1992–2003. *International Journal of Health Geographics* 14(1): 1.

Kohler, B.A., Sherman, R.L., Howlader, N. et al. (2015). Annual report to the nation on the status of cancer, 1975–2011, featuring incidence of breast cancer subtypes by race/ethnicity, poverty, and state. *Journal of National Cancer Institute* 107(6): 1–25.

Koplan, J.P. (2002). The small world of global health. *Mount Sinai Journal of Medicine* 69: 291–298.

Kruger, D.J., Brady, J.S., Shirey, L.A.(2008). Using GIS to facilitate community-based public health planning of diabetes intervention efforts. *Health Promotion Practice* 9: 76–81.

Kuczmarski, R.J., Ogden, C.L., Grummer-Strawn, L.M., Flegal, K.M., Guo, S.S., Wei, R., and Johnson, C.L. (2000). CDC growth charts: United States. *Advance Data* (314): 1–27.

Kuczmarski, R.J., Ogden, C.L., Guo, S.S., Grummer-Strawn, L.M., Flegal, K.M., Mei, Z., and Johnson, C.L. (2002). 2000 CDC Growth Charts for the United States: Methods and development. *Vital and Health Statistics. Series 11, Data from the National Health Survey* (246): 1–190.

Kwate, N.O.A., Yau, C.Y., Loh, J.M., and Williams, D. (2009). Inequality in obesogenic environments: Fast food density in New York City. *Health Place* 15: 364–373.

Lake, A. and Townshend, T. (2006). Obesogenic environments: Exploring the built and food environments. *The Journal of the Royal Society for the Promotion of Health* 126(6): 262–267.

Larson, N.I., Story, M.T., and Nelson, M.C. (2009). Neighborhood environments: Disparities in access to healthy foods in the US. *American Journal of Preventive Medicine* 36(1): 74–81.

Li, F., Fisher, K.J., Brownson, R.C., and Bosworth, M. (2005). Multilevel modelling of built environment characteristics related to neighborhood walking activity in older adults. *Journal of Epidemiology and Community Health* 59: 558–564.

Loomis, B.R., Kim, A.E., Goetz, J.L., and Juster, H.R. (2013). Density of tobacco retailers and its association with sociodemographic characteristics of communities across New York. *Public Health* 127(4): 333–338.

Macintyre, S., Ellaway, A., and Cummins, S. (2002). Place effects on health: How can we conceptualize, operationalize and measure them? *Social Science & Medicine* 55(1): 125–139.

Mainous, A.G.I. and Hueston, W.J. (1997). Using other people's data: The ins and outs of secondary data analysis. *Family Medicine* 29: 568–571.

Mason, T.J. (1975). Atlas of cancer mortality for US Counties, 1950–1969. *British Journal of Cancer* 33(2): 236–237.

Matthews, S.A., Moudon, A.V., and Daniel, M. (2009). Work group II: Using Geographic Information Systems for enhancing research relevant to policy on diet, physical activity, and weight. *American Journal of Preventive Medicine* 36(4): S171–S176.

Mendis, S., Armstrong, T., Bettcher, D., Branca, F., Lauer, J., Mace, C., and Tang, K.C. (2015). Global status report on noncommunicable diseases 2014. Geneva, Switzerland: World Health Organization.

Miller, B.A., Hankey, B.F., and Thomas, T.L. (2002). Impact of sociodemographic factors, hormone receptor status, and tumor grade on ethnic differences in tumor stage and size for breast cancer in US women. *American Journal of Epidemiology* 155: 534–545.

Misra, R. and Lager, J. (2009). Ethnic and gender differences in psychological factors, glycemic control and quality of life among adult type 2 diabetes patients. *Journal of Diabetes and Its Complications* 23: 54–64.

Morris, G.J., Mitchell, E.P. (2008). Higher incidence of aggressive breast cancer in African-American women: a review. *Journal of the National Medical Association* 100(6): 698–702.

Moore, B.J. (2003). Supersized America: Help your patients regain control of their weight: Identifying fast food as a contributor to overeating. *Cleveland Clinic Journal of Medicine* 70(3): 237–240.

Mota, J., Almedia, M., Santos, P., and Ribeiro, J.C. (2005). Perceived neighborhood environments and physical activity in adolescents. *Preventive Medicine* 41(5–6): 834–836.

Mujahid, M.S., Diez Roux, A.V., Morenoff, J.D., and Raghunathan, T. (2007). Assessing the measurement properties of neighborhood scales: from psychometrics to ecometrics. *American Journal of Epidemiology* 165: 858–867.

Mullis, R.M., Blair, S.N., Aronne, L.J. et al. (2004). American Heart Association. Prevention conference VII: Obesity, a worldwide epidemic related to heart disease and stroke: Group IV: Prevention/treatment. *Circulation* 110(18): 484–488.

National Cancer Institute (NCI) and Center for Disease Control (CDC). (2015). The state cancer profiles. http://statecancerprofiles.cancer.gov/about/. Accessed November 27, 2015.

Nelson, A.R., Smedley, B.D., and Stith, A.Y. (Eds.). (2002). *Unequal Treatment: Confronting Racial and Ethnic Disparities in Health Care* (Full Printed Version). Washington, DC: National Academies Press.

Newman, L.A. (2005). Breast cancer in African-American women. *Oncologist* 10: 1–14.

Ng, M., Fleming, T., Robinson, M. et al. (2013). Global, regional, and national prevalence of overweight and obesity in children and adults during 1980–2013: A systematic analysis for the global burden of disease study 2013. *The Lancet* 384(9945): 766–781.

Nicoll, L.H. and Beyea, S.C. (1999). Using secondary data analysis for nursing research. *Association of Operating Room Nurses Journal* 69(2): 428–433.

Oakes, J.M., Mâsse, L.C., and Messer, L.C. (2009). Work group III: Methodologic issues in research on the food and physical activity environments: Addressing data complexity. *American Journal of Preventive Medicine* 36(4): S177–S181.

Obarzanek, E. and Pratt, C.A. (2002). Girls health Enrichment Multi-site Studies (GEMS): New approaches to obesity prevention among young African-American girls. *Ethnicity & Disease* 13(1 Suppl. 1): S1–S5.

Ogden, C.L., Carroll, M.D., Curtin, L.R., Lamb, M.M., and Flegal, K.M. (2010). Prevalence of high body mass index in US children and adolescents, 2007–2008. *Journal of the American Medical Association* 303(3): 242–249.

Ogden, C.L. and Flegal, K.M. (2010). Changes in terminology for childhood overweight and obesity. *AGE* 12: 12.

Onis, M.D., Blössner, M., and Borghi, E. (2010). Global prevalence and trends of overweight and obesity among preschool children. *American Journal of Clinical Nutrition* 92: 1257–1264.

Papas, M.A., Alberg, A.J., Ewing, R., Helzlsouer, K.J., Gary, T.L., and Klassen, A.C. (2007). The built environment and obesity. *Epidemiologic Reviews* 29(1): 129–143.

Paquet, C., Daniel, M., Kestens, Y., Leger, K., and Gauvin, L. (2008). Field validation of listings of food stores and commercial physical activity establishments from secondary data. *International Journal of Behavioral Nutrition and Physical Activity* 5: 58.

Peek, M.E., Cargill, A., and Huang, E.S. (2007). Diabetes health disparities a systematic review of health care interventions. *Medical Care Research and Review* 64(5 Suppl.): 101S–156S.

Pickle, L.W., Szczur, M., Lewis, D.R., and Stinchcomb, D.G. (2006). The crossroads of GIS and health information: A workshop on developing a research agenda to improve cancer control. *International Journal of Health Geographics* 5(1): 51.

Pikora, T.J., Bull, F.C., Jamrozik, K., Knuiman, M., Giles-Corti, B., and Donovan, R.J. (2002). Developing a reliable audit instrument to measure the physical environment for physical activity. *American Journal of Preventive Medicine* 23: 187–194.

Powell, L.M., Slater, S., Mirtcheva, D., Bao, Y., and Chaloupka, F. (2007). Food store availability and neighborhood characteristics in the United States. *Preventive Medicine* 44: 189–195.

Raudenbush, S.W. and Sampson, R.J. (1999). Ecometrics: Toward a science of assessing ecological settings, with application to the systematic social observation of neighborhoods. *Sociological Methodology* 29: 1–41.

Robertson, G.L., Lebowitz, M.D., O'Rourke, M.K., Gordon, S., and Moschandreas, D. (1998). The National Human Exposure Assessment Survey (NHEXAS) study in Arizona—Introduction and preliminary results. *Journal of Exposure Analysis and Environmental Epidemiology* 9(5): 427–434.

Robertson-Wilson, J. and Giles-Corti, B. (2010). Walkability, neighborhood design and obesity. In A.A. Lake, T.G. Townshend, and S. Alvanides (Eds.), *Obesogenic Environments: Complexities, Perceptions and Objective Measures*. London, U.K.: Wiley-Blackwell.

Saelens, B.E. and Glanz, K. (2009). Work group I: Measures of the food and physical activity environment: Instruments. *American Journal of Preventive Medicine* 36(4): S166–S170.

Shaw, J.E., Sicree, R.A., and Zimmet, P.Z. (2010). Global estimates of the prevalence of diabetes for 2010 and 2030. *Diabetes Research and Clinical Practice* 87(1): 4–14.

Shen, W., Punyanitya, M., Chen, J., Gallagher, D., Albu, J., Pi-Sunyer, X., Lewis, C.E., Grunfeld, C., Heshka, S., and Heymsfield, S.B. (2006). Waist circumference correlates with metabolic syndrome indicators better than percentage fat. *Obesity* 14(4): 727–736.

Signorello, L.B., Schlundt, D.G., Cohen, S.S., Steinwandel, M.D., Buchowski, M.S., McLaughlin, J.K., and Blot, W.J. (2007). Comparing diabetes prevalence between African Americans and Whites of similar socioeconomic status. *American Journal of Public Health* 97(12): 2260–2267.

Simmons, E. (2013). Rural obesity in the United States: Causes, consequences, and a need for change. Capstone Project. Lexington, VA: Washington and Lee University. 1–30. http://hdl.handle.net/11021/24168. Accessed April 6, 2014.

Simon, M.S. and Severson, R.K. (1997). Racial differences in breast cancer survival: The interaction of socioeconomic status and tumor biology. *American Journal of Obstetrics & Gynecology* 176: S233–S239.

Singh, G.K., Kogan, M.D., and Van Dyck, P.C. (2008). A multilevel analysis of state and regional disparities in childhood and adolescent obesity in the United States. *Journal of Community Health* 33(2): 90–102.

Skinner, A.C. and Skelton, J.A. (2014). Prevalence and trends in obesity and severe obesity among children in the United States, 1999–2012. *JAMA Pediatrics* 168(6): 561–566.

Spiegel, A.M. and Alving, B.M. (2005). Executive summary of the Strategic Plan for National Institutes of Health Obesity Research. *American Journal of Clinical Nutrition* 82(Suppl.): 211S–214S.

Stewart, A.F. (2008). The United States endocrinology workforce: a supply-demand mismatch. The Journal of Clinical Endocrinology and Metabolism 93: 1164–6.

Stopher, P., FitzGerald, C., and Zhang, J. (2008). Search for a global positioning system device to measure person travel. *Transportation Research Part C* 16: 350–369.

Story, M., Giles-Corti, B., Yaroch, A.L., Cummins, S., Frank, L.D., Huang, T.T., and Lewis, L.B. (2009). Work group IV: Future directions for measures of the food and physical activity environments. *American Journal of Preventive Medicine* 36: S182–S188.

Tang, T.S., Ayala, G.X., Cherrington, A., Rana, G. (2011). A review of volunteer-based peer support interventions in diabetes. *Diabetes Spectrum* 24: 85–98.

Thomson, C.S., Hole, D.J., Twelves, C.J., Brewster, D.H., and Black, R.J. (2001). Prognostic factors in women with breast cancer: Distribution by socioeconomic status and effect on differences in survival. *Journal of Epidemiology and Community Health* 55(5): 308–315.

Thornton, L.E., Pearce, J.R., and Kavanagh, A.M. (2011). Using Geographic Information Systems (GIS) to assess the role of the built environment in influencing obesity: A glossary. *International Journal of Behavioral Nutrition and Physical Act* 8(71).

Timperio, A., Crawford, D., Telford, A., and Salmon, J. (2004). Perceptions about the local neighborhood and walking and cycling among children. *Preventive Medicine* 38(1): 39–47.

Troped, P.J., Saunders, R.P., Pate, R.R., Reininger, B., Ureda, J.R., and Thompson, S.J. (2001). Associations between self-reported and objective physical environmental factors and use of a community rail-trail. *Preventive Medicine* 32: 191–200.

Trust for America's Health. (2011). How obesity threatens America's future 2011. Washington, DC: Robert Wood Johnson Foundation. http://healthyamericans. org/reports/obesity2010/. Accessed June 1, 2011.

USDA. (2015). Defines food deserts. Washington, DC: American Nutrition Association.

Velasquez-Mieyer, P., Neira, C.P., Nieto, R., and Cowan, P.A. (2007). Obesity and cardiometabolic syndrome in children. *Therapeutic Advances in Cardiovascular Diseases* 1(1): 61–81.

Ward, E., Jemal, A., Cokkinides, V., Singh, G.K., Cardinez, C., Ghafoor, A., and Thun, M. (2004). Cancer disparities by race/ethnicity and socioeconomic status. *CA: A Cancer Journal for Clinicians* 54: 78–93.

Whitlock, E.P., Williams, S.B., Gold, R., Smith, P.R., and Shipman, S.A. (2005). Screening and interventions for childhood overweight: A summary of evidence for the US prevention service task force. *Pediatrics* 116(1): 125–144.

WHO cross-national study on health behaviors in school-aged children from 28 countries: Findings from the United States. (2000). *Journal of School Health* 70(6): 227–228.

YRBS. (2007). Youth risk behavior survey questionnaire. ftp://ftp.cdc.gov/pub/data/ yrbs/2015/2015_hs_questionnaire.pdf. Accessed February 15, 2015.

3

Birth Health

Geographic information systems (GIS) provide powerful methodological tools that have been used to confirm the neighborhood impact on adverse birth outcomes (Li et al. 2010). Neighborhoods that suffer high burdens of birth morbidity (e.g., low birth weight, birth defects) may have more in common than just individual maternal risk factors. Susceptibility to morbidity may also be a function of maternal environment. The literature has strongly advocated the use of GIS in surveillance of the maternal environment and its impact on birth morbidity (Stallones et al. 1992; Reif et al. 1996; Richards et al. 1999; Ali et al. 2002; Dangendorf et al. 2002; Cromley 2003; Croner 2003; Elgethun et al. 2003; Cockings et al. 2004). Spatial analysis of birth morbidity provides a way to identify inadequate health-care access as well as potential environmental and behavioral causal factors. Cluster analysis is frequently used to identify an unusually high occurrence of birth morbidity that is clustered in space and time (Alexander and Cuzick 1992). The most commonly used spatial analytical methods in birth morbidity studies are probability mapping (Waller and Gotway 2004), spatial filtering (Talbot et al. 2000; Rushton 2003), Bayesian smoothing (Johnson 2004), and cluster detection methods such as SaTScan (Forand et al. 2002; Ozdenerol et al. 2005). These methods not only map spatial variability of birth health but also attempt to correct for rate instability and preserve case confidentiality (National Vital Statistics Reports 2007). In probability mapping, the statistical significance of rates is mapped (Waller and Gotway 2004). Spatial filtering generates simulated rates in comparison to the actual rate to reveal significantly high rates (Rushton 2003). The Bayesian smoothing process pulls rates toward a regional or a national rate, making rates more stable (Langford 1994). Cluster detection methods evaluate the overall tendency toward clustering (Forand et al. 2002). Other geographic studies explore neighborhood risk characteristics and evaluate neighborhood effects on birth health (Diez Roux 2001, 2004; Picket and Pearl 2001; Macintyre et al. 2002; Kawachi and Berkman 2003; Oaks 2004; Grady 2006).

The purpose of this chapter is to introduce data sets related to birth morbidity and highlight issues that affect data integration in GIS with detailed discussion of the spatial patterns and characteristics of birth health such as low birth weight, birth defects, and infant mortality. From these, spatial patterns meaning may be derived about their underlying mechanisms including individual and local environmental risk factors and access to health-care services.

Birth Morbidity Data

Birth morbidity data can be obtained from electronic birth certificate files from state health departments' division of vital records and statistics in the United States. Some states have pregnancy risk assessment monitoring systems. These population-based surveillance systems randomly sample women from the state birth records and collect information on experiences before, during, and shortly after pregnancy. Most countries have either existing surveillance systems or special birth morbidity studies that monitor and collect data. European countries have population-based registry of congenital malformations (birth defects) as part of the Eurocat EU program (Calzolari et al. 2007). In developing countries, birth defects monitoring in the exposed populations has been adopted as a short-term tool to assess health risks associated with waste management options (WHO 2007; Signorelli et al. 2008). Hospital discharge data are also used in birth outcomes research, usually obtainable at the zip code level of geography. They can be used for special studies such as birth defects but generally are inadequate for population-based studies of birth morbidity because they are restricted to patients treated in the hospitals. Many exposure studies relevant to birth outcomes draw their populations from birth records or birth defects registries (Xiang et al. 2000; Bell et al. 2001; Meyer et al. 2006; Rull et al. 2006), population-based disease studies (Ward et al. 2000; Ritz and Costello 2006), and local clinics such as Women, Infants, and Children clinics (Koch et al. 2002).

Most countries have major national surveys administered to a representative sample of the country's population that include some maternal health-related questions along with nutritional assessment, chronic conditions, injuries, health service utilization such as National Health and Nutrition Examination Survey and the National Health Interview Survey of the United States. Unlike standard vital statistics and morbidity data, local surveys focus on maternal health-related issues that present a broader perspective on maternal health than just the presence of disease and deal with a diverse array of maternal health-related topics such as nutritional status, mental health, stress, and family and neighborhood characteristics. Respondents' residential address is included in surveys but that information is not routinely released.

In the United States, vital records of birth data offer both maternal and infant health indicators. Birth records document a wide range of conditions that affect newborn infants, such as birth weight, gestational weeks (the number of weeks in the gestation period), congenital malformations, obstetric procedures, and abnormal conditions. Mother's marital status, age, race, education, prenatal visits (the number of total visits to a physician during pregnancy), birth multiple (the number of births the mother gave), alcohol use (number of drinks consumed per week), and tobacco use (the number of cigarettes smoked per day) are also included in the birth records. The primary limitation of the birth records is the inherent quality of data for the mother. Recall of timing and number of events during pregnancy such as tobacco use, alcohol use, and perinatal visits have errors and inconsistencies. More disaggregate data on cigarette smoking, drinking, dietary factors, and family history are needed. Additional information such as mother's income, nativity (country of origin), and exposure to environmental pollution could be collected by existing systems and improve the quality of collected data.

Mother's residential address is the geographic identifier on birth records. GIS's geocoding functionality converts the tabular data files to a consistent geographical structure. Geocoding is the mechanism that identifies locations on a map with real coordinates such as addresses. The lists of addresses with birth records are automatically matched against a digital geographic file of streets and addresses through GIS software. Incorrect, incomplete, or ambiguous address information such as P.O. Box and rural route style addresses reduces the matching confidence. This requires manual intervention such as interactively editing the addresses if mistakes exist.

When case (birth record) and address are matched, census geographies such as census tract, block group, and blocks from the geographic files are transferred to the case records and data linkages to any of these geographic areas could then occur. Case and population counts for census areas and/or zip codes are obtained after they are spatially joined in GIS. This process allows birth outcome distributions to be studied in relation to ecologically associative environmental and socioeconomic status factors. The use of GIS has been found to provide valuable insights into potential relationships between environmental exposures and adverse birth outcomes. Though the relevant exposure may have occurred some place other than the mother's residence, the current address may not represent the environment before and during pregnancy or prior to death. Mothers live their lives in overlapping environments that include, for example, their residence, work, and school. These may all be within the same neighborhood or not. It is unknown how long the identified residence on birth records was the mother's residence. Toxic neighborhood exposures are not necessarily stable over time. The conditions that lead to birth defects or infant mortality can result from lifelong exposures and behaviors.

Low Birth Weight

Infants born prematurely (i.e., infants born less than 37 weeks gestation), with low birth weight (i.e., infants born less than 2500 g), and/or with congenital defects are at increased risk of death because of undeveloped or poorly developed organs and the inability to physiologically respond to their external environment (Grady and Enander 2009). Therefore, the proportion of newborns weighing less than 2500 g is a major concern among researchers, health professionals, and social planners in any community. Widespread differences in mean birth weight have often been observed among various racial and ethnic groups and they are commonly attributed, at least in part, to genetic causes (Alexander 1985). Although low birth weight is associated with a wide range of biological and social factors, there is also evidence that other less well-known environmental factors may affect the growing fetus (Kramer et al. 1992). Several reproductive outcomes, such as birth weight, premature births, and fertility, are frequently thought to be affected by exposure to complex chemical mixtures (Viena and Polan 1984; Deane et al. 1989; Bove et al. 1992). GIS is used to estimate populations for census-driven boundaries (i.e., census tracts and block groups) and view that information (i.e., demographic, socioeconomic) in map format along with the residential address of mothers and maternal health characteristics. This allows to adjust for difference in maternal age distributions and study rates (i.e., low birth weight) in small geographic units that may cross town lines but have similar point source of pollution, socioeconomic, and environmental characteristics.

Using GIS to study the spatial characteristics of low birth weight is complicated by the lack of methodological guidance and the scarcity of accurate integrated spatial-morbidity databases (Kafadar 1996; Brody et al. 2002; Maantay 2002; McElroy et al. 2003; Rushton 2003). A representative, but not exhaustive, set of methods exists for analyzing spatial clusters of birth morbidity with individual point data or aggregated data that still maintain the stability of the estimated rates by constructing a continuous smoothed map (Rushton and Lolonis 1996; Kulldorf 1997; Talbot et al. 2000). The infant low birth weight rate is a ratio between the low birth weight infants in a geographical area and the total births in the same area over some period of time. Infant low birth weight rates are calculated for different geographic scales such as census tracts, block groups, and zip codes. Results can vary significantly depending upon the level of geographic scale that is analyzed (Schneider et al. 1993; Choi et al. 2003; Diem 2003). For example, spatial patterns of birth morbidity vary in relation to area-based census tract, block group, and zip code level measures of socioeconomic status (Richards et al. 1999). Additionally, aggregation bias represents an inherent source of error for these analyses (Bonner et al. 2003). Aggregation bias results from the rather arbitrary means by which GIS aggregate individual cases at some geographic organizational unit such as a census tract. The coarser the spatial scale the higher the potential for aggregation bias.

Estimation methods differ in how one aggregates individual level data. Consequently, aggregation bias may manifest itself distinctively in one method versus the other. There is a need for an exploratory, integrative, and multiscalar approach to assessing geographic patterns of birth outcomes, since different methods identify different patterns. A methodological comparison can provide insight about the limitations and benefits of varying approaches when mapping birth morbidity. Ozdenerol et al. (2005) demonstrated how SaTScan and the spatial filtering technique better reflect spatial variation when individual point data are used. They suggest that both estimation methods provide a useful way to characterize the spatial aspects of low birth weight. The results from both software can easily be incorporated into GIS. SaTScan results provide the radius, latitude and longitude coordinates, and the P-value for the most likely clusters in a database. On the other hand, the Distance Mapping and Analysis Program produces morbidity rates using spatial filters and tests for significance using Monte Carlo simulations. Its results are isarithmic maps that exhibit a continuous spatial distribution. Since both methods perform better on the point data, there is no need to include any analyses of aggregated data. The Bernoulli model was chosen by Ozdenerol et al. (2005), which requires information about the location of a set of cases and controls. Customarily, low birth weight geocoded points are selected as cases and normal birth weight geocoded points as the controls.

The spatial filtering method has been used to study clusters of congenital anomalies, infant mortality, and other forms of birth morbidity (Rushton et al. 1995; Kafadar 1996; Talbot et al. 2000). The spatial filtering method works well with both aggregated data and individual point data. Once estimated rates are assigned to each grid point, isarithmic maps are constructed in GIS. The isarithmic maps have many advantages in comparison with other conventional thematic maps that provide an indication of the level of a disease by area. The resultant clusters are not constrained by the borders of geographic units, and sudden transitions between levels of two neighboring areas are avoided (Burkom 2003). An interpolation technique (i.e., the inverse-distance weighing) is used to construct the isarithmic maps. Since inverse-distance weighing represents the average of the values of the surrounding points, weighed by the inverse of the distance to those points, the process is based on the assumption of positive spatial autocorrelation depicting a continuous gradient exists between points in a linear way (Burkom 2003). It is assumed that the probability of a birth resulting in a low birth weight is equal to the proportion of all births in the region that resulted in low birth weight (Rushton et al. 1995).

The SaTScan method has been used to study clusters of sudden infant death syndrome and congenital anomalies (Kulldorf 2001; Cromley and McLafferty 2002; Gregorio and Samociuk 2003). SaTScan uses a circular window of different sizes that scans the study area until a certain percent (e.g., half) of the total population is included. This circle is the most probable cluster and has a rate that is the least likely to happen by chance alone. SaTScan also accounts for multiple testing through the calculation of the highest likelihood of occurrence for all possible cluster locations and sizes

(Census 2000; Kulldorf 2001; Cromley and McLafferty 2002). Although a range of probabilities can be displayed using SaTScan, only the most highly significant estimates are displayed.

SaTScan accounts for multiple testing of the highest likelihood of occurrence for all possible cluster locations and sizes while spatial filtering does not. In addition, there may be somewhat of a "ceiling effect" with SaTScan. This maximum value ensures that the detected clusters, regardless of their location and size, are clusters detected without any preselection bias. The maximum allowed value of a spatial cluster does not mean that one has to prespecify the size of a cluster before running an analysis. It simply means the largest allowed cluster would contain 50% of the at-risk population in the study area. This maximum value is reasonable because a cluster is expected to concentrate in certain areas of the study region. If a cluster covers most of the study area, then the location and size of the study area is no longer meaningful in that study area. Consequently, SaTScan clusters have an inherent but adjustable "cap" on cluster size, whereas spatial filtering is somewhat less limited. The tendency or capacity of spatial filtering to yield clusters with considerably more geographic variability than SaTScan raises the issue of sample generalizability. The potential for greater variability of the characteristics of births within a spatial filtering cluster may provide some analytic benefits.

Based on Ozdenerol and her colleagues' findings, the spatial filtering technique provides many advantages over SaTScan (Ozdenerol et al. 2005). The SaTScan-detected cluster is of a noncontinuous circular shape, which often conceals spatial pattern—even though the actual geographic coordinates of each case and control are used through the Bernoulli method. Spatial filtering, on the other hand, treats low birth weight rates as a continuous spatial distribution. Spatial filtering provides estimated risks for low birth weight incidence for each location in the map while SaTScan provides the statistical significance of the likely clusters after adjusting for multiple testing. Spatial filtering calculates risk estimates while using predetermined circle sizes defined either by geographical or "constant or near constant population size rather than constant geographic size" (Kulldorf 1997). SaTScan, on the other hand, uses circles of different sizes when searching over a grid (Cromley 2003). The size of clusters identified by spatial filtering depends on the identified filter size, such as the radius of the circles. Figure 3.1 reveals that progressively larger spatial filters removed local spatial variability, which eventually produced an approximate uniform pattern of low birth weight. Clusters increased in size and additional clusters emerged toward the northeastern part of the county when the 0.5 mile filter size is used (Figure 3.2). As the filter size radius increased to 0.6 mile, the localized clusters unioned to a larger uniform pattern covering the western portion of the county (Figure 3.3). Once a larger spatial filter size of 0.8 mile is applied, the spatial filtering technique lost the ability to detect elevated rates except for the most densely populated areas.

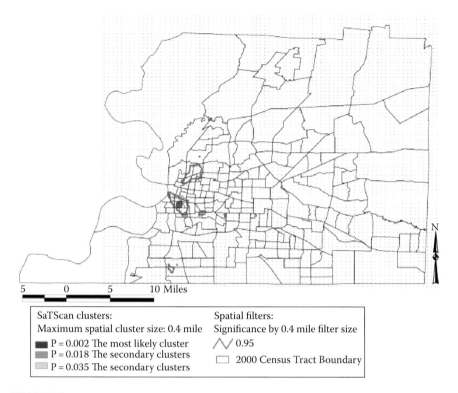

FIGURE 3.1
Areas with statistically significant high rates of low birth weights, Shelby County, Tennessee, 2000–2002. The maps show SaTScan clusters with a maximum spatial cluster size of 0.4 miles. It also shows significant spatial filter clusters with a maximum 0.4 mile filter size.

Infant Mortality

Infant mortality data are generated from death certificates. These data include demographic characteristics of the infant, information about the cause of death, and contributing factors. There are errors and inconsistencies in diagnosis and assigning causes when multiple causes are present. There are two address data in the death certificates. One is the place of death, which could be a hospital, nursing home, or other health-care facility. This information can be used in the utilization of health-care services, especially for populations in which racial and ethnic disparities in health and health-care use exist. The other address is the residence of the decedent, which is used to better understand the contextual environmental factors that hinder or facilitate health.

Individual risk factors are frequently poor predictors of infant mortality (Gould et al. 2003). Patterns exist above and beyond individual characteristics.

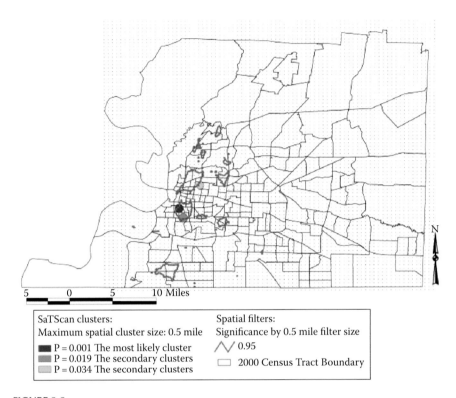

FIGURE 3.2
Areas with statistically significant high rates of low birth weights, Shelby County, Tennessee, 2000–2002. The maps show SaTScan clusters with a maximum spatial cluster size of 0.5 miles. It also shows significant spatial filter clusters with a maximum 0.5 mile filter size.

Neighborhood context is recognized as an important risk factor for infant mortality (Metcalfe et al. 2011). Infant mortality is often disproportionately prevalent in disadvantaged neighborhoods. Disadvantaged neighborhoods share much in common including unemployment, substandard housing, crime, inadequate health care, lack of transportation, residential segregation, poor social support, and inadequate culture and family structure. Furthermore, they frequently contain disproportionately high prevalence of point sources of pollutants (Chakraborty 2001) and suffer environmental hazards that are unique to their area (Northridge et al. 2003). Infants born in these neighborhoods share common risk factors for mortality such as prematurity and low birth weight (Wise et al. 1994; Ashton 2006). Overall patterns of racial disparities in infant mortality and secular changes in rates of prematurity as well as birth weight patterns in infants of African immigrant populations contradict the genetic theory of race and point toward social and environmental mechanisms (David and Collins 2007). Foreign-born Mexican American women possess maternal risk factors similar to that

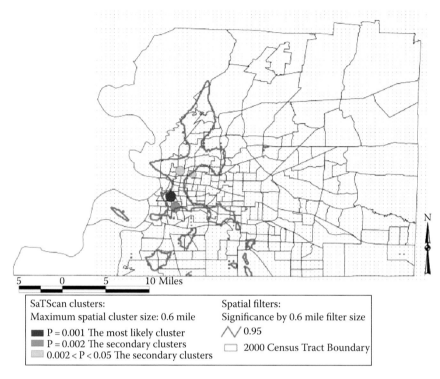

FIGURE 3.3
Areas with statistically significant high rates of low birth weights, Shelby County, Tennessee, 2000–2002. The maps show SaTScan clusters with a maximum spatial cluster size of 0.6 miles. It also shows significant spatial filter clusters with a maximum 0.6 mile filter size.

of ethnic minority groups such as African immigrant women but demonstrate lower rates of infant mortality (Palloni and Marenoff 2001; Gould et al. 2003). Violent crime characterized as a neighborhood attribute was positively associated with preterm birth and low birth weight among non-Hispanic white and black women (Messer et al. 2006). Low socioeconomic status and racial and ethnic minority status are frequently confounded with high exposure to environmental pollutants and also associated with reduced access to high-quality prenatal care (Williams et al. 2007).

Although there are clearly "hot spots" in which the residents are at greater risk for poverty, violence, and infant mortality, investigations that adequately control for the interrelatedness of demographic, economic, environmental, and individual risk factors are still rare. Individual risk factors are usually analyzed together with variables generated from GIS (e.g., levels of contamination, census demographics). Individual-level birth and infant death records are commonly grouped to calculate rates and ratios by aggregating to a census level of geography (e.g., census tract, zip codes).

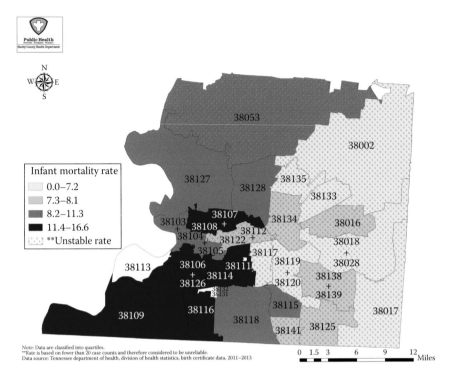

FIGURE 3.4
Infant mortality rates by zip code in Shelby County, Tennessee, 2011–2013. (Courtesy of Memphis and Shelby County Health Department, Memphis, TN.)

Figure 3.4 shows an example map of infant mortality rate by zip code level of geography. Due to suspected heterogeneities (nonuniform patterns) in neighborhood effects, descriptive analysis is usually performed for each level of aggregation, and odds ratios and 95% confidence intervals are calculated for different race groups (e.g., whites and blacks) separately at each neighborhood aggregation level. Models are also adjusted for maternal age and other maternal characteristics. The spatial patterns derived from this analysis may include aggregation bias due to the differential distribution of confounding variables created by grouping (Cromley and McLafferty 2002). With grouped data, that census unit becomes the unit of observation. Observations are assumed independent. However, results may be biased (Bryk and Raudenbush 1992). Ecological fallacy occurs when analyses based on grouped data lead to conclusions different from those based on individuals. Individual-based approaches do not take into account that individuals are nested within their local environments. This nested structure implies that the observations based on individuals sharing the same local environment are not statistically independent. Using multilevel models helps to decompose these rates and ratios by studying individuals nested within their environments. Hierarchical linear modeling (HLM) considers this nested structure. Several studies have demonstrated how HLM can be used to analyze the relationships between neighborhood-level environmental data, individual risk factors, and birth outcomes (Ahern et al. 2003; Kaufman et al. 2003, Williams et al. 2007).

HLM is specifically designed to facilitate the analysis of interactions between level of the individual (level 1) and local environment (level 2). In simple terms, infant and neighborhood levels. Variables specific to each birth (e.g., maternal age) are called level-1 variable. Variables that are derived for each census tract and all infants born within the same tract share the same value (e.g., poverty level) are called level-2 variable. Using the combined level-1 and level-2 data sets, HLM provides a way to analyze the degree to which the effects of individual level variables in birth data vary systematically or randomly as a function of level-2 variables. For example, Williams et al. (2007) assessed the impact of the local environment on birth outcomes and with HLM, they could determine the effects of potentially remediable environmental conditions after controlling for individual factors (Williams et al. 2007). They found that ambient concentrations of criterion air pollutants (e.g., lead and sulfur dioxide) demonstrated a sizeable negative effect on birth weight while the economic characteristics of the mother's residential census tract (e.g., poverty level) also negatively influenced birth weight.

GIS complements HLM in the area of birth outcomes by linking infant data to aggregate environmental, demographic, and economic neighborhood-level data. GIS is highly utilized for area-based measures (e.g., census administrative boundaries, zip codes) and proximity-based measures within varying proximities of residential location (e.g., buffers) to investigate neighborhood effect (i.e., deprivation). GIS-based birth outcome studies can focus both upon the broad trends across geography and outcomes and individual estimates and their neighborhood effect, though there are limitations related to neighborhood-effect research (Elo et al. 2009; Vinikor-Imler et al. 2011). Constructing area boundaries could lead to spurious associations in neighborhood effects. The probable units of effect prior to data collection, variable construction, and analysis should be identified through statistically significant findings. For example, walking is likely a local (not census tract) behavior and might be associated with more block-group-level poverty (Laraia et al. 2007). Census-defined neighborhoods such as census tracts and block groups provide comparable information and continue to have utility for neighborhood-effect research (Messer et al. 2012). In the United States, the range of exposures used to represent relevant neighborhood-level effects have been largely limited to sociodemographic variables available from the census (Laraia et al. 2006). Census tracts are relatively homogeneous with respect to population characteristics (average about 4000 inhabitants), socioeconomic status, and housing/living conditions. Census block groups are smaller geographic units contained within a census tract. In the United Kingdom, the Townsend Material Deprivation Score and the Carstairs Deprivation Index have been widely utilized as established area-level indices for the comparison of deprivation effects across a variety of geographic regions (Caughy et al. 2001).

Another limitation is the problem of endogeneity, whereby unobserved variables most likely related to exposures and outcomes must be considered and needs further exploration. Neighborhoods shape the culture of its residents while the residents within the neighborhood construct the culture of the neighborhood. They function as both markers and sources of exposure. For example, a neighborhood's walkability may not be directly related to weight gain or smoking during pregnancy. Rather, observed associations may result from external processes such as selection bias or from an unmeasured confounding factor. Simply, nonsmokers could be attracted to walkable neighborhoods or social norms enforced by residents of walkable neighborhoods could encourage physical activity. It is difficult to determine whether "poor places make people poor, or poor places attract people" (Tienda 1991). Therefore, causation is difficult to establish (Greenland 1988; Walter 1991). The conclusion one would draw out of neighborhood-effect research suggests there are markers of health-promoting or -damaging environments that are associated with birth health and mother's behavior.

In addition to neighborhood-level investigations, GIS is used to compute continuous spatial distribution of infant mortality rates (Rushton et al. 1995). There are a number of methods for spatial interpolation of point data in the GIS environment (Lam and De Cola 1993). Rushton and Lolonis (1996) used a method known as "punctual kriging," a widely used method in the geosciences (Oliver et al. 1989; Carrat and Valleron 1992; Webster et al. 1994). Taking any location at random, they defined its infant mortality rate by selecting an arbitrarily sized region surrounding it of a size sufficient to capture enough observations of births and deaths to estimate the rate. They believed if they repeated this for a grid of such estimates, they could interpolate the infant mortality rate as a continuous spatial distribution. From each location on a square grid with approximately 0.4 miles apart, a circle of 0.4 mile radius was drawn and the number of births and infant deaths within the circle area was computed. Figure 3.5 shows overlapping circular zones generated around grid points (Rushton and Lolonis 1996). The infant mortality rate was defined for the grid location at the center of each circular area as the ratio of infant deaths to births found in the area for the time period specified. They computed a continuous spatial distribution of infant mortality rates by a contouring procedure (Openshaw et al. 1987, 1988; Turnbull et al. 1990). These probabilities, portrayed as isarithmic maps, show areas that have significantly high rates of infant mortality.

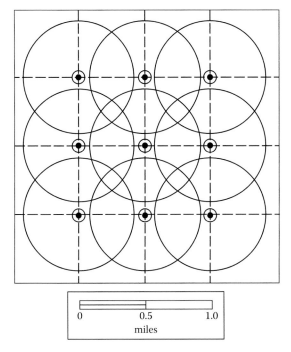

FIGURE 3.5
Overlapping circular zones generated around grid points in the Rushton and Lolonis method of analyzing clusters. (From Rushton, G. and Lolonis, P.: Exploratory spatial analysis of birth defect rates in an urban population. *Stat. Med.* 1996. 15(7–9). 717–726. Copyright Wiley-VCH Verlag GmbH & Co. KGaA. Reproduced with permission.)

Birth Defects

Birth defects have been studied as a possible effect of toxic pollutants from point sources such as hazardous waste sites. This is because such effects can be monitored with either existing surveillance systems or special studies. Upton et al. (1989) state that selection of an appropriate disease outcome for surveillance at hazardous waste sites is difficult. When the exposed population is small, it is difficult to detect rare outcomes. The complex chemical mixtures found at hazardous waste sites do not usually result in clearly definable health outcomes. The identification of exposed individuals and the estimation of compounds, dose, duration, and frequency of exposure are problematic in assessing health outcomes. Although such difficulties exist, several GIS-based studies have evaluated exposures that might increase the risk of birth defects (Cordier et al. 2004; Viel et al. 2008; Vinceti et al. 2009).

GIS have been used to develop composite databases of the sources of environmental contaminants, describe the source of contaminants, and evaluate the contamination zones around these sources (Osleeb and Kahn 1998). Geographic and physical descriptions of the source, information on the rate of release into the atmosphere, the surface and groundwater hydrology, the geologic and physical characteristics of receiving land, and air are evaluated in chemical dispersion models (Chakraborty and Armstrong 1995). These models incorporate GIS for spatial data compilation, analysis, mapping spatial variability, and visualization at a variety of spatial scales (Wagenet and Hutson 1996). GIS helps to determine how the contamination travels by aerial drift or through groundwater to affect adjacent areas as well. The output dispersal footprint from the dispersion model is transferred into a GIS map and footprint area is outlined and overlaid with census and birth morbidity data to map the exposed populations and model the exposure risk. The automated environment of the GIS enables one to evaluate the sensitivity of the exposure estimates.

In an Italian study (Vinceti et al. 2009), a GIS was used for a population-based case–control investigation near a municipal solid waste incinerator. GIS was incorporated into a dispersion model for exposure assessment and for geographical localization of cases and controls. Based on the Gaussian analytic solution of the turbulent diffusion equation (WindDimula 2008), the model attempts to estimate concentration levels of exposure to incinerator emissions of dioxins and furans and simulates both short-term and climatological concentrations of calm wind conditions that are usual in the city of Reggio Emilia in northern Italy (Cirillo and Poli 1992). Figure 3.6 shows the output concentration levels of exposure, given in a raster format on regular grids.

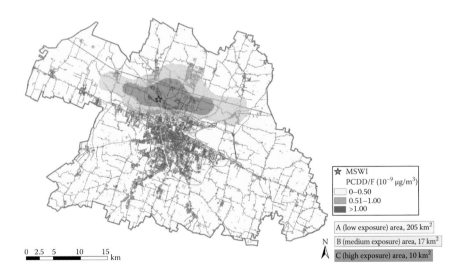

FIGURE 3.6
Map of exposure to polychlorinated dibenzo-p-dioxins and dibenzofurans in the city of Reggio Emilia in northern Italy, around the municipal solid waste incinerator. (Courtesy of Vinceti, M. et al., *Int. J. Health Geogr.*, 8, 8, 2009.)

Cases of congenital anomalies in the offspring or in aborted fetuses of women were obtained from a population-based registry of congenital malformations as part of the Eurocat EU program (Calzolari et al. 2007). A mother was considered exposed when her address was comprised within the intermediate and high (B and C) exposure areas, after inputting it in the GIS. They found an excess prevalence of chromosomal anomalies in the middle exposure area, which is difficult to interpret since the risk was not increased in the high exposure area, and no association was found in the two epidemiologic studies that specifically examined this category of birth defects (Cresswell et al. 2003; Dummer et al. 2003; Cordier et al. 2004). GIS-based studies are considered useful to further monitor findings.

A number of studies have also used remote sensing technologies in addition to GIS to examine birth defects as a result of agricultural pesticide exposure. Remote sensing provided crop classification. The classified crop patterns and the percentage of total acreage by crop type that was treated with specific pesticides were mapped in GIS. Crop production patterns around mothers' residences and proximity to the agricultural fields are used as surrogates for mother's exposure to agricultural pesticides during pregnancy. GIS's spatial functions such as distance measurement, buffering, and overlay analysis are used to measure residential proximity to agricultural fields (Lu et al. 2000; Bell et al. 2001; Royster et al. 2002), estimate crop acreages within specified buffer zones (Schreinemachers 2003), and link pesticide use to land use variables (Ritz and Costello 2006; Rull et al. 2006) and are used in household-level collection of environmental and biological samples (Allpress et al. 2008). GIS-based exposure modeling is used to model probabilities of pesticide use (Ward et al. 2000, 2006; Xiang et al. 2000; Meyer et al. 2006) and examine historical patterns of pesticide use (Brody et al. 2002, 2004).

In addition to exposure assessment studies, GIS is used to study spatial variability in birth defects (Rushton and Lolonis 1996). Rushton and Lolonis have found areas that have significantly high birth defect rates for the period between 1983 and 1990 in Des Moines, Iowa. They computed birth defect rates for a regular lattice of grid points arbitrarily located at approximately half-mile intervals throughout the Des Moines area. Birth defect rates, based on these lattice points were represented as a continuous distribution by interpolating values between the grid points. They presented their result by an isarithmic map with simple contouring procedures using GIS.

Conclusion

GIS analysis helps view the maternal health problems at a variety of geographic scales, from individual to the local neighborhood scale and to the global scale. Understanding birth health from a geographic perspective leads to better prevention and intervention and improves access to health services. The types of methods used in analyzing birth outcomes depend fundamentally on access to birth outcome data (i.e., area or point). Complete and up-to-date birth defect registries and vital records should be established and maintained and integrated with GIS so that birth outcome and maternal health-related statistics can be attached to their geography (Ebener et al. 2015). Integration of geography in any health facility survey would also contribute to facilitating the use of GIS. Collecting maternal health indicator data at different types of administrative divisions and different periods of time makes application of GIS very challenging. Some procedures also emphasize the arrangement of birth events, not just in relation to population at risk, but also in relation to potential sources of contamination. Finding and developing accurate GIS data and environmental data for this purpose is often time consuming, expensive, and challenging.

References

Ahern, J., Pickett, K.E., Selvin, S., and Abrams, B. (2003). Preterm birth among African American and white women: A multilevel analysis of socioeconomic characteristics and cigarette smoking. *Journal of Epidemiology & Community Health* 57(8): 606–611.

Alexander, F.E. and Cuzick, J. (1992). Methods for the assessment of disease clusters. In P. Elliott, J. Cuzick, D. English, and R. Stern (Eds.), *Geographical & Environmental Epidemiology: Methods for Small-Area Studies*. Oxford, U.K.: Oxford University Press, pp. 238–247.

Alexander, G.R. (1985). Racial differences in the relation of birthweight. *Public Health Report* 5: 39–47.

Ali, M., Emch, M., Donnay, J.P., Yunus, M., and Sack, R.B. (2002). Identifying environmental risk factors for endemic cholera: A raster GIS approach. *Health Place* 8(3): 201–210.

Allpress, J.L.E., Curry, R.J., Hanchette, C.L., Philips, M.J., and Wilcosky, T.C. (2008). A GIS-based method for household recruitment in a prospective pesticide exposure study. *International Journal of Health Geographics* 7: 18.

Ashton, D. (2006). Prematurity infant mortality: The scourge remains. *Ethnicity and Disease* 16(2 Suppl. 3): S3-S58-S62.

Bell, E.M., Hertz-Picciotto, I., and Beaumont, J.J. (2001). Case-cohort analysis of agricultural pesticide applications near maternal residence and selected cases of fetal death. *American Journal of Epidemiology* 154: 702–710.

Bonner, M.R., Han, D., Nie, J., Rogerson, P., Vena, J.E., and Freudenheim, J.L. (2003). Positional accuracy of geocoded addresses in epidemiologic research. *Epidemiology* 14(4): 408–412.

Bove, F.J., Fulcomer, M.C., and Klotz, J.B. (1992). Public drinking water contamination and birth weight, fetal deaths and birth defects: A cross sectional study. Trenton, NJ: New Jersey Department of Health.

Brody, J., Aschengrau, A., McKelvey, W., Rudel, R., Swartz, C., and Kennedy, T. (2004). Breast cancer risk and historical exposure to pesticides from wide-area applications assessed with GIS. *Environmental Health Perspective* 112: 889–897.

Brody, J.G., Vorhees, D.J., Melly, S.J., Swedis, S.R., Drivas, P.J., and Rudel, R.A. (2002). Using GIS and historical records to reconstruct residential exposure to large-scale pesticide application. *Journal of Exposure Analysis and Environmental Epidemiology* 12(1): 64–80.

Bryk, A.S. and Raudenbush, S.W. (1992). *Hierarchical Linear Models: Applications and Data Analysis Methods*. Newsbury Park, CA: Sage.

Burkom, H.S. (2003). Bio surveillance applying scan statistics with multiple, disparate data sources. *Journal of Urban Health* (1 Suppl. 1): 57–65.

Calzolari, E., Pierini, A., Astolfi, G., Bianchi, F., Neville, A.J., and Rivieri, F. (2007). Associated anomalies in multi-malformed infants with cleft lip and palate: An epidemiologic study of nearly 6 million births in 23 EUROCAT registries. *American Journal of Medical Genetics A* 143: 528–537.

Carrat, F. and Valleron, A.J. (1992). Epidemiological mapping using the "Kriging" method: Application to an influenza-like illness epidemic in France. *American Journal of Epidemiology* 135(11): 1293–1300.

Caughy, M.O., O'Campo, P.J., and Patterson, J. (2001). A brief observational measure for urban neighborhoods. *Health and Place* 7: 225–236.

Census, U.S. (2000). Census 2000 gateway. http://www.census.gov/main/www/cen2000.html. Accessed April 3, 2013.

Chakraborty, J. (2001). Acute exposure to extremely hazardous substances: An analysis of environmental equity. *Risk Analysis* 21(5): 883–895.

Chakraborty, J. and Armstrong, M.P. (1995). Using geographic plume analysis to assess community vulnerability to hazardous accidents. *Computers, Environment, and Urban Systems* 19(5–6): 1–17.

Choi, K.M., Serre, M.L., and Christakos, G. (2003). Efficient mapping of California mortality fields at different spatial scales. *Journal of Exposure Analysis and Environmental Epidemiology* 13(2): 120–133.

Cirillo, M. and Poli, A. (1992). An intercomparison of semiempirical diffusion models under low wind speed, stable conditions. *Atmospheric Environment* 26A: 765–774.

Cockings, S., Dunn, C.E., Bhopal, R.S., and Walker, D.R. (2004). Users' perspectives on epidemiological, GIS and point pattern approaches to analyzing environment and health data. *Health Place* 10(2): 169–182.

Cordier, S., Chevrier, C., Robert-Gnansia, E., Lorente, C., Brula, P., and Hours, M. (2004). Risk of congenital anomalies in the vicinity of municipal solid waste incinerators. *Occupational and Environmental Medicine* 61: 8–15.

Cresswell, P.A., Scott, J.E., Pattenden, S., and Vrijheid, M. (2003). Risk of congenital anomalies near the Byker waste combustion plant. *Journal of Public Health Medicine* 25: 237–242.

Cromley, E.K. (2003). GIS and disease. *Annual Review of Public Health* 24: 7–24.

Cromley, E.K. and McLafferty, S.L. (2002). *GIS and Public Health*. New York: The Guilford Press, pp. 8–9.

Croner, C.M. (2003). Public health, GIS, and the internet. *Annual Review of Public Health* 24: 57–82.

Dangendorf, F., Herbst, S., Reintjes, R., and Kistemann, T. (2002). Spatial patterns of diarrhoeal illnesses with regard to water supply structures—A GIS analysis. *International Journal of Hygiene and Environmental Health* 205(3): 183–191.

David, R. and Collins, J. (2007). Maternal health and infant health in diverse set-
    tings: Disparities in infant mortality: What's genetics got to do with it? *American
    Journal of Public Health* 97(7): 1191–1197.
Deane, M., Swan, S.H., Harris, J.A., and Neutra, R.R. (1989). Adverse pregnancy
    outcomes in relation to water contamination, Santa Clara County, California.
    American Journal of Epidemiology 129: 894–904.
Diem, J.E. (2003). A critical examination of ozone mapping from a spatial-scale per-
    spective. *Environmental Pollution* 125(3): 369–383.
Diez Roux, A.V. (2001). Investigating neighborhood and area effects on health.
    American Journal of Public Health 91(11): 783–1789.
Diez Roux, A.V. (2004). Estimating neighborhood effects: The challenges of causal
    inference in a complex world. *Social Science and Medicine* 58(10): 1953–1960.
Dummer, T.J., Dickinson, H.O., and Parker, L. (2003). Adverse pregnancy outcomes
    around incinerators and crematoriums in Cumbria, North West England, 1956–93.
    Journal of Epidemiology and Community Health 57: 456–461.
Ebener, S., Guerra-Arias, M., Campbell, J., Tatern, A.J., Moran, A.C., Johnson, F.A.,
    Fogstad, H. et al. (2015). The geography of maternal and new born health: The
    state of the art. *International Journal of Health Geographics* 14: 19.
Elgethun, K., Fenske, R.A., Yost, M.G., and Palcisko, G.J. (2003). Time-location analy-
    sis for exposure assessment studies of children using a novel global positioning
    system instrument. *Environmental Health Perspective* 111(1): 115–122.
Elo, I.T., Culhane, J.F., Kohler, I.V., O'Campo, P., Burke, J.G., Messer, L.C., Kaufman,
    J.S., Laraia, B.A., Eyster, J., and Holzman, C. (2009). Neighborhood depriva-
    tion and small-for-gestational-age term births in the United States. *Pediatric and
    Perinatal Epidemiology* 23: 87–96.
Forand, S.P., Talbot, T.O., Druschel, C., and Cross, P.K. (2002). Data quality and the
    spatial analysis of disease rates: Congenital malformations in New York State.
    Health and Place 8: 191–199.
Gould, J.B., Madan, A., Qin, C., and Chavez, G. (2003). Perinatal outcomes in two
    dissimilar immigrant populations in the United States: A dual epidemiologic
    paradox. *Pediatrics* 111 (6 Part 1): e676–e682.
Grady, S.C. (2006). Racial disparities in low birthweight and the contribution of
    residential segregation: A multilevel analysis. *Social Science and Medicine* 63:
    3013–3029.
Grady, S.C. and Enander, H. (2009). Geographic analysis of low birthweight and infant
    mortality in Michigan using automated zoning methodology. *International
    Journal of Health Geographics* 8: 10. doi:10.1186/1476-072X-8-10.
Greenland, S.R.J. (1988). Conceptual problems in the definition and interpretation of
    attributable fractions. *American Journal of Epidemiology* 128: 1185–1197.
Gregorio, D.I. and Samociuk, H. (2003). Breast cancer surveillance using gridded
    population units, Connecticut, 1992–1995. *Annals of Epidemiology* 13: 42–49.
Johnson, G.D. (2004). Small area mapping of prostate cancer incidence in New York
    State (USA) using fully Bayesian hierarchical modeling. *International Journal of
    Health Geographics* 3: 29.
Kafadar, K. (1996). Smoothing geographical data, particularly rates of disease.
    Statistics in Medicine 15(23): 2539–2560.
Kaufman, J.S., Dole, N., Savitz, D.A., and Herring, A.H. (2003). Modeling community
    effects on preterm birth. *Annals of Epidemiology* 13(5): 377–384.

Kawachi, I. and Berkman, L.F. (2003). *Neighborhoods and Health*. New York: Oxford University Press.

Koch, D., Lu, C., Fisker-Anderson, J., Jolley, L., and Fenske, R. (2002). Temporal association of children's pesticide exposure and agricultural spraying: Report of a longitudinal biological monitoring study. *Environmental Health Perspective* 110: 829–833.

Kramer, M.D., Lynch, C., Isacson, P., and Hanson, J.W. (1992). The association of waterborne trihalomethane exposure with intrauterine growth retardation. Birth Defects Registry Report. Iowa City, IA: Department of Preventive Medicine, University of Iowa.

Kulldorff, M. (1997). A spatial scan statistic. *Communications in Statistics: Theory and Methods* 26: 1481–1496.

Kulldorff, M. (2001). Prospective time periodic geographical disease surveillance using a scan statistic. *Journal of the Royal Statistical Society [Series A]* 164(1): 61–72.

Lam, N.S.-N. and De Cola, L. (Eds.). (1993). *Fractals in Geography*. Englewood Cliffs, NJ: Prentice Hall.

Langford, I. (1994). Using empirical Bayes estimates in the geographical analysis of disease risk. *Area* 26(2): 142–149.

Laraia, B., Messer, L., Evenson, K., and Kaufman, J.S. (2007). Neighborhood factors associated with physical activity and adequacy of weight gain during pregnancy. *Journal of Urban Health* 84(6): 793–806.

Laraia, B.A., Messer, L.C., Kaufman, J.S., Dole, N., Caughy, M., O'Campo, P., and Savitz, D.S. (2006). Direct observation of neighborhood attributes in an urban area of the U.S. south. *International Journal of Health Geographics* 5: 11.

Li, Q., Kirby, R.S., Sigler, R.T., Hwang, S., Lagory, M.E., and Goldenberg, R.L. (2010). A multilevel analysis of intimate partner violence among low-income pregnant women in Jefferson County, Alabama. *American Journal of Public Health* 100(3): 531–539.

Lu, C., Fenske, R.A., Simcox, N.J., and Kalman, D. (2000). Pesticide exposure of children in an agricultural community: Evidence of household proximity to farmland and take home exposure pathways. *Environmental Research* 84: 290–302.

Maantay, J. (2002). Mapping environmental injustices: Pitfalls and potential of geographic information systems in assessing environmental health and equity. *Environmental Health Perspective* 110(Suppl. 2): 161–171.

Macintyre, S., Ellaway, A., and Cummins, S. (2002). Place effects on health: How can we conceptualize, operationalize and measure them? *Social Science and Medicine* 55(1): 125–139.

McElroy, J.A., Remington, P.L., Trentham-Dietz, A., Robert, S.A., and Newcomb, P.A. (2003). Geocoding addresses from a large population-based study: Lessons learned. *Epidemiology* 14(4): 399–407.

Messer, L.C., Kaufman, J.S., Dole, N., Herring, A., and Larain, B.A. (2006). Violent crime exposure classification and adverse birth outcomes: A geographically defined cohort study. *International Journal of Health Geographics* 5: 22. doi:10.1186/1476-072X-5-22.

Messer, L.C., Vinikoor-Imler, L.C., and Laraia, B.A. (2012). Conceptualizing neighborhood space: Consistency and variation of associations for neighborhood factors and pregnancy health across multiple neighborhood units. *Health Place* 18: 801–813.

62

Metcalfe, A., Lail, P., Ghali, W.A., and Sauve, R.S. (2011). The association between neighborhoods and adverse birth outcomes: A systematic review and meta-analysis of multi-level studies. *Peadiatric and Perinatal Epidemiology* 25: 236–245.

Meyer, K.J., Reif, J.S., Veeramachaneni, D.N.R., Luben, T.J., Mosley, B.S., and Nuckols, J.R. (2006). Agricultural pesticide use and hypospadias in eastern Arkansas. *Environmental Health Perspective* 114(10): 1589–1595.

National Vital Statistics Reports. (2007). Births: Final data for 2005. *Centers for Disease Control and Prevention* 56(6): 98–103.

Northridge, M.E., Stover, G.N., Rosenthal, J.E., and Sherard, D. (2003). Environmental equity and health: Understanding complexity and moving forward. *American Journal of Public Health* 93(2): 209–214.

Oaks, J.M. (2004). The (mis)estimation of neighborhood effects: Causal inference for a practicable social epidemiology. *Social Science and Medicine* 58: 1929–1952.

Oliver, M., Webster, R., and Gerrard, J. (1989). Geostatistics in physical geography. Part I: Theory. *Transactions of the Institute of British Geographers* 14: 259–269.

Openshaw, S., Chartton, M., and Craft, A.W. (1988). Searching for leukemia clusters using a geographic analysis machine. *Papers in Regional Science* 64: 95–106.

Openshaw, S., Chartton, M., Wymer, C., and Craft, A.W. (1987). A mark I geographic analysis machine for the automated analysis of point data sets. *International Journal of Geographic Information Systems* 1: 335–358.

Osleeb, J.P. and Kahn, S. (1998). Integration of geographic information. In V.H. Dale and M.R. English (Eds.), *Tools to Aid Environmental Decision Making*. New York: Springer-Verlag, pp. 161–189.

Ozdenerol, E., Williams, B.L., Kang, S.Y., and Magsumbol, M.S. (2005). Comparison of spatial scan statistic and spatial filtering in estimating low birth weight clusters. *International Journal of Health Geographics* 4: 19.

Palloni, A. and Morenoff, J.D. (2001). Interpreting the paradoxical in the Hispanic paradox. *Annals of the New York Academy of Sciences* 954(1): 140–174.

Picket, K. and Pearl, M. (2001). Multilevel analyses of neighborhood socioeconomic context and health outcomes: A critical review. *Journal of Epidemiology and Community Health* 55(2): 111–122.

Reif, J.S., Keefe, T., and Stallones, L. (1996). Reproductive, neurobehavioral, and other disorders in communities surrounding the Rocky Mountain Arsenal. Government Reports Announcements & Index (GRA&I). Atlanta, GA: U.S. Department of Human Health and Service.

Richards, T.B., Croner, C.M., Rushton, G., Brown, C.K., and Fowler, L. (1999). Information technology: Geographic information systems and public health: Mapping the future. *Public Health Reports* 114(4): 359.

Ritz, B. and Costello, S. (2006). Geographic model and biomarker-derived measures of pesticide exposure and Parkinson's disease. *Annals of the New York Academy of Sciences* 1076(1): 378–387.

Royster, M.O., Hilborn, E.D., Barr, D., Carty, C.L., Rhoney, S., and Walsh, D. (2002). A pilot study of global positioning system/geographical information system measurement of residential proximity to agricultural fields and urinary organophosphate metabolite concentrations in toddlers. *Journal of Exposure Analysis and Environmental Epidemiology* 12(6): 433–440.

Rull, R.P., Ritz, B., and Shaw, G.M. (2006). Neural tube defects and maternal residential proximity to agricultural pesticide applications. *American Journal of Epidemiology* 163(8): 743–753.

Rushton, G. (2003). Public health, GIS, and spatial analytic tools. *Annual Review of Public Health* 24(1): 43–56.

Rushton, G., Krishnamurti, D., Krishnamurthy, R., and Song, H. (1995). Spatial analysis of infant mortality rates in Des Moines, Iowa, 1989–1992. In *Public Health Conference on Records and Statistics and the National Committee on Vital and Health Statistics*, Washington, DC.

Rushton, G. and Lolonis, P. (1996). Exploratory spatial analysis of birth defect rates in an urban population. *Statistics in Medicine* 15(7–9): 717–726.

Schneider, D., Greenberg, M.R., Donaldson, M.H., and Choi, D. (1993). Cancer clusters: The importance of monitoring multiple geographic scales. *Social Science & Medicine* 37(6): 753–759.

Schreinemachers, D.M. (2003). Birth malformations and other adverse perinatal outcomes in four U.S. wheat-producing states. *Environmental Health Perspective* 111(9): 1259–1264.

Signorelli, C., Riccò, M., and Vinceti, M. (2008). Waste incinerators and human health: A state-of-the-art review. *Annali di igiene* 20: 251–277.

Stallones, L., Nuckols, J., and Berry, J. (1992). Surveillance around hazardous waste sites: Geographic information systems and reproductive outcomes. *Environmental Research* 59(1): 81–92.

Talbot, T.O., Kulldorff, M., Forand, S.P., and Haley, V.B. (2000). Evaluation of spatial filters to create smoothed maps of health data. *Statistics in Medicine* 19(Part 17/18): 2399–2408.

Tienda, M. (1991). Poor people and poor places: Deciphering effects on poverty outcomes. In J. Huber (Ed.), *Macro-Micro Linkages in Sociology*. Newbury Park, CA: Sage, pp. 244–263.

Turnbull, W., Iwano, E.J., Burnett, W.S., Howe, H.L., and Clark, L.C. (1990). Monitoring for clusters of disease: Application to leukemia incidence in upstate New York. *American Journal of Epidemiology* 132: S136–S143.

Upton, A.C., Kneip, T., and Toniolo, P. (1989). Public health aspects of toxic chemical disposal sites. *Annual Review of Public Health* 10: 1–25.

Viel, J.F., Clement, M.C., Hagi, M., Grandjean, S., Challier, B., and Danzon, A. (2008). Dioxin emissions from a municipal solid waste incinerator and risk of invasive breast cancer: A population-based case-control study with GIS-derived exposure. *International Journal of Health Geographics* 7: 4.

Vienna, N.J. and Polan, A.K. (1984). Incidence of low birth weight among Love Canal residents. *Science* 226: 1217–1219.

Vinceti, M., Malagoli, C., Fabbi, S., Teggi, S., Rodolfi, S., Garavelli, L., Astolfi, G., and Rivieri, F. (2009). Risk of congenital anomalies around a municipal solid waste incinerator: A GIS-based case-control study. *International Journal of Health Geographics* 8: 8.

Vinikoor-Imler, L.C., Messer, L.C., Evenson, K.R., and Laraia, B.A. (2011). Neighborhood conditions are associated with maternal health behaviors and pregnancy outcomes. *Social Science & Medicine* 73(9): 1302–1311.

Wagenet, R.J. and Hutson, J.L. (1996). Scale-dependency of solute transport modeling/GIS applications. *Journal of Environmental Quality* 25(3): 499–510.

Waller, L.A. and Gotway, C.A. (2004). *Applied Spatial Statistics for Public Health Data*, Vol. 368. New York: John Wiley & Sons.

Walter, S.D. (1991). The ecologic method in the study of environmental health. II. Methodologic issues and feasibility. *Environmental Health Perspective* 94: 67.

Ward, M.H., Lubin, J., Giglierano, J., Colt, J.S., Wolter, C., Bekiroglu, N., and Nuckols, J.R. (2006). Proximity to crops and residential exposure to agricultural herbicides in Iowa. *Environmental Health Perspective* 114(6): 893–897.

Ward, M.H., Nuckols, J.R., Weigel, S.J., Maxwell, S.K., Cantor, K.P., and Miller, R.S. (2000). Identifying populations potentially exposed to agricultural pesticides using remote sensing and a geographic information system. *Environmental Health Perspective* 108: 5–12.

Webster, R., Oliver, M.A., Muir, K.R., and Mann, J.R. (1994). Kriging the local risk of a rare disease from a register of diagnoses. *Geographic Analysis* 26: 168–185.

Williams, B.L., Pennock-Roman, M., Suen, H.K., Magsumbol, M.S., and Ozdenerol, E. (2007). Assessing the impact of the local environment on birth outcomes: A case for HLM. *Journal of Exposure Science and Environmental Epidemiology* 17: 1–13.

WindDimula 3.0.6. (2008). Maind s.r.l. http://www.maind.it/software/windimula.htm. Accessed May 5, 2011.

Wise, P.H., Wampler, N., and Barfield, W. (1994). The importance of extreme prematurity and low birthweight to US neonatal mortality patterns: Implications for prenatal care and women's health. *Journal of the American Medical Women's Association (1972)* 50(5): 152–155.

World Health Organization. (2007). Population health and waste management: Scientific data and policy options. Report of a WHO *Workshop*, Rome, Italy, March 29–30, 2007.

Xiang, H., Nuckols, J.R., and Stallones, L. (2000). A geographic information assessment of birth weight and crop production patterns around mother's residence. *Environmental Research* 82(2): 160–167.

4

Infectious Diseases

Infectious diseases move through time and space. They are caused by living microorganisms (e.g., microbes, germs, agents, pathogens) and therefore transmittable directly from person to person and/or via a vector from host to host. Understanding the spatial dynamics of infectious diseases requires an understanding of the spatial determinants of each of these organisms and their interactions. Cases of infectious diseases that spread directly from person to person are not independent given that the outcomes from subjects who live close to each other tend to be positively dependent. Geographic information system (GIS) is a vital tool in investigating this spatial dependence, spatial causation, and spread of new or reemerging infectious diseases.

Since the spread of microbes is not contained by country borders, the threat of global epidemics has serious consequences for international community. Global surveillance of infectious diseases helps in the prevention and control of their spread. Such awareness is essential in promoting future global public health surveillance and coordination among the different international players. Individual countries are responsible for disease surveillance and response. The International Health Regulations require countries that identify public health events of international concern (e.g., cholera, internationally quarantinable disease) to report to the World Health Organization, which disseminates the information, as needed, to other countries (Rodier 2007). Surveillance in low-resource countries are more challenging than surveillance in industrialized countries and that tilts the geographic dimensions of infectious disease spread.

Understanding the mode of transmission and dynamics of transmission risk is critical in any GIS-based assessment of infectious diseases. The spatial distributions reflect the environmental conditions that affect risk and susceptibility and the interactions that facilitate transmission. For vector-borne diseases, the vector, usually an insect, is involved in the transmission of the infectious disease. A reservoir host maintains the pathogen that carries the disease. GIS can be used to analyze the spatial distribution of vector and host populations and the land use and human activity patterns where reservoir host–vector–human interaction is present.

The infectious diseases that spread directly from person to person are called "nonvectored" diseases such as HIV/AIDS, tuberculosis, measles, and influenza, among others. They could be airborne and spread through the process of respiration (i.e., influenza, tuberculosis, or the common cold). Waterborne transmission is one of the major routes of infection.

Waterborne infectious diseases are acquired by drinking water contaminated at its source or in the water distribution system or by direct contact with environmental and recreational waters. Some infectious diseases are caused by toxins from food or soil, such as tetanus. Sexually transmitted diseases (STDs) such as HIV/AIDS and syphilis are spread directly through skin or sexual contact.

A GIS is a perfect platform to integrate infectious disease epidemic models (that focus on identifying changes in susceptible and infected populations, the transmission rate, and incubation periods) with diffusion models and include the space–time dimension of disease spread (Thomas 1992). There are four types of infectious disease diffusion pattern: contagious diffusion, a gradual outward expansion following transport commuting flows, social interactions; hierarchical diffusion, disease spread via the urban hierarchy jumping over long distances from large metropolitan areas to medium- and smaller-sized cities (Wallace and Wallace 1998); network diffusion, disease spread through the geographical and social structuring of human interactions; and mixed pattern, a mix of contagious, hierarchical, and network diffusion (Gould 1995).

Although address information is the key geographical identifier to incorporate infectious disease data into GIS, geographical investigations are often done at aggregated level to predefined geographical units (e.g., zip codes, census tracts, or counties) because of privacy and confidentiality concerns. Key data sources for infectious disease reporting include sentinel systems (reporting systems for cases of notifiable conditions), surveillance systems (e.g., hospital, school, and vector/host zoonotic disease surveillance), and state and local public health surveys. The data sources will be revisited in detail in the next sections of this chapter.

This chapter examines the use of GIS in analyzing infectious diseases. For this purpose, infectious diseases are categorized as vector-borne and nonvectored diseases. This chapter gives the breadth of GIS applications and approaches to show the applicability of spatial analysis for vector-borne and nonvectored diseases and future implications.

Vector-Borne Diseases

All vector-borne zoonotic diseases require the co-occurrence of a pathogen, a vector, one or more reservoir hosts, and the human victim. The most common vectors are mosquitoes and ticks. Dengue fever, malaria, West Nile virus (WNV), and yellow fever are examples of mosquito-borne diseases. Lyme disease (LD), Rocky Mountain spotted fever, Heartland virus, Colorado tick fever, and Southern tick-associated rash illness are examples of tick-borne diseases.

The rapid global diffusion of vector-borne diseases argues for a better understanding of their geographic extent. The number of ecological studies that have used GIS and spatial modeling has increased significantly over the past decade, examining climate factors in addition to biological and ecological determinants. Moreover, with the availability of vector geographic data and local response involving GIS and remote sensing (RS) technologies, particularly in tick, bird, and mosquito surveillance, comparability across studies is enhanced. The purpose of this section is to review GIS applications in two key vector-borne diseases: LD and WNV.

West Nile Virus (WNV)

To predict and explain the spread of WNV, one needs to understand the complex interrelationships among human, avian, and mosquito habitat systems and the interplay of environmental, built environment, and anthropogenic risk factors that influence these systems (Ruiz et al. 2004; Messina et al. 2011; Cooke et al. 2006; Gibbs et al. 2006). Both prospective and retrospective techniques are needed to identify WNV exposure areas (Brownstein et al. 2004). Prospective techniques are mainly early warning approaches based on information on a single component of the WNV transmission cycle, that is, either infected birds or human cases. Retrospective techniques are spatial statistical modeling approaches, relating the incidence cases of either infected dead birds or human cases to a range of risk factors. It is vital to incorporate the spatial–temporal information of the three components of the transmission cycle affecting virus spread—birds (reservoir), mosquitoes (vector), and humans (host)—in the model, to understand the viral transmission. Birds are the main reservoir hosts of WNV, and mosquitoes are the main vector for the virus transmission from birds to humans, horses, and other birds. WNV epidemics mainly occur in summer and autumn in temperate, subtropical, and tropical areas. As WNV is highly influenced by regular, seasonal climate, and environmental changes, it is particularly amenable to spatial and temporal analysis (Epp et al. 2009). Habitat studies of environmental risk exposure need to be validated by models using follow-up data on the distribution of human cases and vectors; otherwise, modeling environmental risk based on habitat alone could be problematic. The geographical distribution of risk areas (based on environmental conditions) needs to be compared to the distribution of vectors and reservoir hosts and human cases (Mather et al. 1996; Ward et al., 2005).

GIS and RS are useful techniques to model habitats since producing a map of infected vectors or hosts for all locales is too difficult. RS models have been used to define suitable landscape features and meteorological variables for vector transmission and potential vector and host mechanisms of dispersal in many parts of the world (Mather et al. 1996; Ward 2005; Jacob et al. 2009, 2010; Winters et al. 2010). Environmental data taken from RS, such as vegetation type and density, elevation, slope, hydrology, and soil moisture, have been used to model host risk exposure. Consequent GIS analysis could quantify associations between risk variables and environmentally sampled covariates (Griffith 2005; Ward 2005; Jacob et al. 2009, 2010; Winters et al. 2010; Young and Jensen 2012). Other common data analyses and functions performed using GIS/RS technology are time series overlay analyses of thematic geographic data and spatial intersection analyses, buffer generation and neighborhood analysis, vector-borne grid generation and network analysis, and disease risk surface modeling (Kitron 2000; Rogers and Randolph 2003). GIS are being used for proactive surveillance, prevention, and control measures and to study patterns of environmental factors such as global climate change and their possible impacts on the spread of WNV (Hay et al. 2000; David et al. 2007; Bouden et al. 2008; Jacob et al. 2010). GIS and RS are also employed by the Centers for Disease Control and Prevention (CDC) and the United States Geological Survey (USGS) to prepare interpretive maps showing WNV activity in North America (CDC 2013).

GIS-based studies examining spatial modeling of WNV infection and risk factors published since the New York City outbreak in 1999 indicate a growing geographic spread of the virus and the importance of climatic factors in addition to biological and ecological determinants on WNV occurrence (Ozdenerol et al. 2013). Furthermore, significant associations have been found between WNV infection and some socioeconomic and microclimatic indicators, though these seem to be location dependent. Table 4.1 presents these studies with GIS methods applied, study region and date, and resultant common risk factors with location-dependent ones indicated. Recent finer-scale microclimatic studies (Davis et al. 2005; Cleckner et al. 2011; Kim et al. 2011; Chuang et al. 2012; Thompson et al. 2013) applied at various temporal scales reemphasize the extent of the role and precision of risk factors and the likely influence they are having on the rapidly growing burden of WNV in the world. Surveillance methods that accommodate spatial analysis at various temporal scales have potential to provide understanding of location-dependent risk factors. Fusing high-temporal-resolution and high-spatial-resolution RS imagery can offer higher predictive accuracies for estimating location-dependent microclimatic conditions associated with WNV occurrence. Making use of such newly available RS products improves the quality of surveillance data and can lead to tangible improvements in the prevention and control of WNV.

There is a growing trend for WNV research to move from descriptive toward predictive studies and more exploratory investigations. Public health officials are increasingly challenged to assess the prevalence and to determine the most common risk factors as well as to track their trends over time. Very few studies have made the leap from spatial prediction to risk-based surveillance (Gosselin et al. 2005; David et al. 2007; Soverow et al. 2009). Deviation from use of linear models to more powerful nonparametric models such as computational neural networks that capture nonlinear relationships with greater accuracy have more applicability in prevention and control,

TABLE 4.1

Summary of West Nile Virus Studies with Common Risk Factors

Analysis/Citation	Region/Date	Common Risk Factors
Spatial analysis of human case incidence		
Local Moran's I (Ruiz et al. 2004, Moran 1950)	Chicago, 2002	Less population density,[a] higher percent of old and white residents,[a] poor drainage, mosquito abatement efforts
Ripley's K test (Messina et al. 2011)	Chicago, 2005–2006	Inner suburbs, less densely populated areas,[a] high percent of white residents,[a] post–World War II housing and a higher median population age, smaller elevation ranges, standing water, more vegetated areas
SaTScan, Local Moran's I (Sugumaran et al. 2009, Moran 1950)	United States, 2002–2008	Study focused on hot spots of human case incidence
Hot spot analysis (Liu et al. 2009)	Connecticut, 2000–2005	Urban/suburban areas
Global Moran's I, NDVI (Degroote et al. 2008)	Iowa, 2003–2006	Less population density[a] and rural agricultural areas, drier conditions
SaTScan (Wimberly et al. 2008)	Northern plains, 2003	Rural areas, irrigated land in rural areas
Global Moran's I (Young and Jensen 2012; Moran 1950)	United States, 2012	Temperature and precipitation ranges
Conditional autoregressive model (Brownstein et al. 2004)	United States, 2013	The number of WNV positive mosquito pools
Spatial–temporal analysis of bird species		
Mapping migration routes (Rappole et al. 2000)	North America	Wintering grounds along coastal plains of Georgia, northern Florida
Bird abundance mapping (David et al. 2007)	British Columbia, 1994–2003	Dead corvid density
GLMM (Yiannakouilas et al. 2006)	Alberta, Canada, 2002–2006	The grassland natural region, rural/suburban areas
NND time model (Ghosh and Guha 2010)	Twin Cities, 2002	Densely populated areas,[a] distance to nearest dead bird and pool location
Mahalanobis distance statistics (Liu et al. 2011)	Virginia, 2011	Mean weekly precipitation, moderate percent impervious surface with 21%–40% canopy density
Spatial analysis of horses		
Kriging (Jacob et al. 2010)	Indiana, 2002	High temperatures in August–September months
Proximity analysis (Ward and Sheurman 2008)	Texas, 2002	Proximity of equine cases to human cases in urban populations

(Continued)

TABLE 4.1 (*Continued*)

Summary of West Nile Virus Studies with Common Risk Factors

Analysis/Citation	Region/Date	Common Risk Factors
LULC analysis, SatScan clustering (Leblond et al. 2007)	France	Rice fields, dry bushes, open water, low elevation salted swamps
SaTScan (Lian et al. 2007)	Texas	Study focused on areas of high risk
Spatial modeling of mosquito pools		
Risk mapping (Cooke et al. 2006)	Mississippi	High road density, low stream density and gentle slopes
Mahalanobis distance statistics (Ozdenerol et al. 2008)	Tennessee, 2004	High percentage of black population, low income, high rental occupation, old structures, vacant housing
Spatial sensitivity analysis (Winters et al. 2010)	Colorado, 2003–2007	Study focused on subcounty scale presentation and how WNV disease occurrence influenced by data aggregation
Spatiotemporal spread, risk mapping (Hernandez-Jouer et al. 2013)	Australia, 2013	Predictive risk-zone mapping
Habitat-based studies		
Maximum likelihood unsupervised classification LULC change matrix (Jacob et al. 2009)	Urbana Champaign, IL, 1991–2003	Residential high canopy coverage
Geospatial models based on LULC (Jacob et al. 2009)	Cook County, IL, 2002–2005	Warmer temperature and heavy precipitation, forest and middle-range built environment
Terrain analysis, ISODATA (Jacob et al. 2010)	Tuskegee, AL	Smaller elevation range
Generation of DEM, spatial hydrological modeling, Eigenvector mapping (Jacob et al. 2011)	Trinidad, 2008–2009	Terrain elevation
Shortest distance analysis (Soverow et al. 2009)	17 U.S. states, 2001–2005	Warmer temperatures, elevated humidity, and heavy precipitation
Raster-based mosquito abundance model (Tachiiri et al. 2006)	British Columbia	Study focused on risk-prone areas
LULC analysis (Epp et al. 2009)	Saskatchewan, CA, 2003–2007	Study focused on risk mapping
NDVI analysis, RS-driven spatial analysis (Calistri et al. 2013)	Morocco	Precipitation
Computational neuronetworks (Ghosh and Guha 2010, 2011)	Twin Cities, MN, 2002–2006	Proximity to wetlands

(Continued)

TABLE 4.1 (*Continued*)

Summary of West Nile Virus Studies with Common Risk Factors

Analysis/Citation	Region/Date	Common Risk Factors
RS studies for early warning systems		
ASTER imagery and MODIS (Liu et al. 2011)	N. Virginia	Elevation and urban built-up conditions that negatively correlated with WNV propagation, land surface temperature that positively correlated with viral transmission
ASTER imagery and MODIS (Liu et al. 2008)	Indianapolis	
ASTER imagery and MODIS (Liu et al. 2009)	Chicago	
AMSR-E-derived models (Chuang et al. 2012)	South Dakota	Air temperature and vegetation opacity and surface water fraction
Tasseled-cap transformation (Cleckner et al. 2011)	Coastal Virginia	Study focused on developing a habitat suitability index
AVIRIS (Thompson et al. 2013)	Fresno, CA	Neglected swimming pools
NDWI (Kim et al. 2011)	Atlanta, GA	Neglected swimming pools
Spatial analysis of genetic variation		
Population genetic analysis (Davis et al. 2005; Bertoletti et al. 2007; Snapinn et al. 2007)	United States	Localized environmental conditions
Population genetic analysis (Bertoletti et al. 2008)	Chicago, 2008	Seasonal variations in microclimatic conditions at finer scale
Spatial uncertainty analysis		
Spatial uncertainty analysis, SaTScan (Wey et al. 2009; SaTScan 2013)	South Dakota	Lower ability to geocode Indian reservations

[a] Location dependent.

especially for chemical control vector abatement programs (Ghosh and Guha 2010, 2011). For example, Ozdenerol et al. (2008) developed an innovative GIS-driven, raster-based technique to construct probability risk maps for areas ecologically most suitable for containing WNV-infected mosquito habitats using Mahalanobis distance statistics model (Jenness 2003). They quantitatively described the entire Memphis/Shelby County, Tennessee, landscape in terms of how similar it is to the "ideal" elevation, slope, land use, temperature, precipitation, and vegetation density for infected mosquitoes. They defined "ideal" by the environmental conditions in areas where WNV-infected mosquitoes were actually found in an epidemiologic week. Figure 4.1 shows the quantity and spatial distribution of highly and moderately suitable habitats in the month of August, during which the highest viral activity occurs.

Real-time studies provide reliable predictions but depend on high-quality input data. Many data sets utilized in such analyses are not collected for such

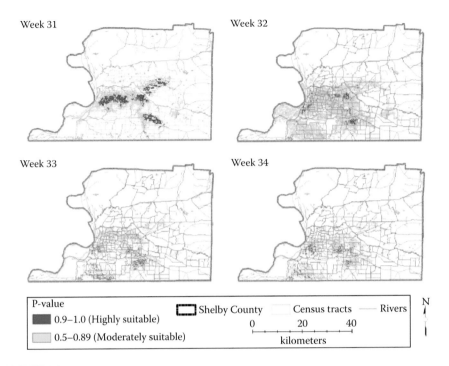

FIGURE 4.1
Habitat suitability based on Mahalanobis distance statistic model (Four weeks of August 2004 in Memphis/Shelby County, Tennessee, United States). This map shows moderately and highly suitable habitats based on P-values. (Courtesy of Ozdenerol, E. et al., *Int. J. Health Geogr.*, 7, 12, 2008.)

spatial analytical purposes and have limitations with respect to spatial coverage, temporal appropriateness, and precision, accuracy of georeferencing and geocoding success. There is only one study published regarding selection bias in the ability to geocode WNV cases (Wey et al. 2009).

Future work lies in strengthening data collection techniques and eliminating some of the inherent spatial uncertainties in this field of research. Studies of phylogeographic analyses (i.e., the study of historical processes responsible for contemporary distributions) need to be further integrated with spatial statistical modeling and accommodate subsequent analysis at various scales. Their applicability in real-world decision-making needs to be investigated, especially in localized environmental conditions.

Regional Variations of WNV Distribution

Overall, spatial epidemiology studies maintained evidence that WNV has become epidemic in North America and those annual outbreaks are likely for the foreseeable future. It has reached Central and South America, though analysis of WNV distribution and public health impacts remain poorly understood in these regions. WNV is known to have reached western

China (Xinjiang province), but there is no evidence of WNV in the rest of China: however, surveillance reports do not exist from most of China, so there is little knowledge about the possible distribution of WNV in China. Reemergence of the virus in India is a threat to highly populated countries of Southeast Asia. Recent confirmed cases in Japan and South Korea prompted these countries to start nationwide WNV surveillance efforts and international collaboration of intense research efforts to determine the particular strains involved.

WNV has been appearing more frequently in the Mediterranean and eastern European nations in recent years. Genome sequence analysis of WNV strains isolated in Italy in 2008 and 2009 showed they were closely related suggesting the virus had overwintered and established an endemic cycle. In 2012, early detection of human cases of WNV disease has been reported also in Sardinia (Anderson et al. 1999), in Greece, Israel, and the occupied Palestinian territory (European Centre for Disease Prevention and Control 2012), and this trend might predict increased viral activity in the Mediterranean area. Figure 4.2 presents the geographic dimension of the current distribution of WNV activity in the world.

The WNV activity in southern Europe suggests that the avian hosts contribute to the movement of the virus. The virus is translocated by the viremic migratory birds to the temperate and tropical regions and more likely persist in the southern wintering sites in the Western Hemisphere (Rappole et al. 2000).

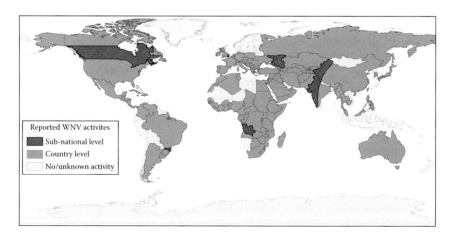

Reported WNV activites
Sub-national level
Country level
No/unknown activity

FIGURE 4.2
The figure presents the geographic dimension of the current distribution of West Nile virus (WNV) activity in the world. The dark hatch pattern represents areas in which WNV activity has been reported at the subnational level in particular regions of some countries and the lighter-gray shading signifies countries with (at least) some reported WNV activity, and the lightest gray represents counties with no or unknown activity. (Courtesy of Ozdenerol, E. et al., *Int. J. Environ. Res. Public Health*, 10(11), 5399, 2013.)

Future GIS-based ecological studies are needed to examine the frequency of cycling of active virus in the avian hosts and detect the favorable habitat and climatic conditions for the migrating vectors. Future studies should also consider whether the local vectors are capable of transmitting the virus.

The prospective techniques are mainly early warning approaches based on information on a single component of the WNV transmission cycle, that is, either infected birds or human cases. Some studies found that positive bird information does not appear to offer much predictive value in spatial models (Yiannakouilas et al. 2006) while others have reported successful prediction of human risk with this approach (Bian and Li 2006). The dynamic distribution of bird populations, sampling bias (due to voluntary public surveillance), and other factors could all complicate any meaningful associations between WNV infection in humans and birds. More vegetation means increased habitats for both WNV vector and bird reservoir hosts, with urban green areas having the necessary tree cover to support bird populations and contact between migratory and residential bird species, which has been found to be important in WNV amplification (Rappole et al. 2000).

Lower overall land-cover diversity may indicate greater concentration of bird species that are efficient hosts for the virus and thus the potential for higher incidence of WNV in humans (Rappole et al. 2000). Future studies on land-cover diversity and avian cases need to be conducted.

Large and dense groups of migratory birds gather in wetlands and potentially reach every part of the southeastern United States, Mexico, Central America, the Caribbean Island, and South America during their migration south to wintering sites and nearly every part of North America during their migration north to breeding sites (Rappole et al. 2000). The suitable wetland sites along the coastal plain of Georgia, northern Florida, and Alabama are likely to receive the largest number of potentially infected hosts and therefore may be the most likely places for future outbreaks, if the necessary ornithophilic mosquitoes are sufficiently active and abundant (Rappole et al. 2000). Future spatial–temporal analyses on intense monitoring of fall and winter avian concentrations for abnormal die-offs are highly suggested in the coastal areas.

More measures on improving the representativeness of bird surveillance data should be taken. Areas with low human population and low bird densities may choose to focus surveillance efforts on mosquito and human cases. Media communications should encourage the public to increase their reporting of dead bird sightings.

Improved surveillance efforts of avian and mosquito species across metropolitan regions as well as prevention controls in reducing breeding grounds will have a significant impact on reducing the risk of WNV in humans. The investigation of reservoir hosts and vectors is the key to determine WNV risk-prone areas. Figure 4.3 shows an example investigation of vectors such as persistent weekly positive mosquito pools for WNV risk-prone areas at the zip code level. The assessment risk of WNV to humans cannot be made outside of

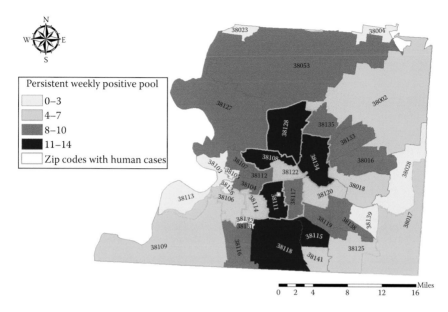

FIGURE 4.3

Persistent weekly positive mosquito pools of West Nile virus (WNV) and zip codes with human cases of WNV in Shelby County, TN, United States. *Note*: Data are classified into optimal breaks using the Jenks Method. (Courtesy of Memphis and Shelby County Health Department, Memphis, TN.)

the context of the urban environment in which it is present. The effect of mosquito control is apparent. Effective public outreach and health education are key factors to eliminate WNV risk to humans. Future research in the response of residents to the risk of WNV and a qualitative assessment of the political and social factors related to mosquito control should also be carried out.

Future research should be concerned with the noticeable change in pattern and magnitude of human WNV outbreaks between years. It is possible that various mosquito control efforts (and lack thereof) on the part of political entities may have an effect on the patterns of human WNV illness, and these impacts have not been studied well. Seasonal climate and temperature patterns and soil moisture characteristics play a part in *Culex* abundance and WNV transmission and should also be explored more thoroughly. Storm water systems also may be particularly suitable for vector production. Catch basins often provide the stagnant water and the cool moist environment needed by *Culex* to survive in hot dry weather and deposit their eggs, and an exploration of catch-basin characteristics and their locations would be a valuable contribution toward increased understanding of WNV transmission. Delineation of WNV transmission cycles at local scales can be used to develop hypotheses related to the disease spread and thus advance our understanding of the complexity of avian–mosquito–human environmental systems on a microscale. The modes of transmission such as blood transfusion need to be revisited through public health studies.

Continued development in satellite RS offers frequent worldwide coverage in high spatial and spectral resolutions appropriate for assessing vegetation and moisture conditions at local and intermediate scales, including the availability of historical images comparable with historical WNV data. RS can be used more often and more effectively in assessing WNV-carrying mosquito habitats over time and space.

More empirical knowledge on the spatiotemporal patterns of mosquitoes is needed. For example, *Culex quinquefasciatus* larval habitats and their underlying contribution to adult population are needed to devise an effective WNV surveillance program in regions of the southern United States and the Caribbean (Jacob et al. 2011).

Habitat-based interventions should emphasize the link between foraging behaviors of egg-laying mosquitoes and the availability of breeding sites in evaluation of environmental management programs.

More research is needed to assess the effects of expected climate shifts on WNV transmission and studies need to be designed on longer time scales. Shorter-time-scale studies also have strengths that include the ability to restrict variables to 1 year or season, which can improve the power to study geographic variation, the age, immunity, status of human cases, socioeconomic characteristics, and so on, linking vector to reservoir host populations, and more.

GIS-based ecological studies have demonstrated how GIS can be used to better understand WNV to track outbreaks and to create prevention and control techniques. Additionally, spatial statistics are also very useful when determining relationships between WNV cases and other variables, such as ecological or social factors. We need to improve our understanding of the extent spatial–temporal methodologies can be applied to other WNV-prone areas of the world, given that the characteristics of transmission cycle and surveillance programs vary significantly from one region to another.

Lyme Disease

LD is the most common tick-borne disease in the temperate zones of the Northern Hemisphere (Ogden et al. 2008). The common pathogen known to cause LD in North America, *Borrelia burgdorferi*, is transmitted from mammal to mammal (small-sized, ground-dwelling vertebrate hosts) by ticks of the genus *Ixodes scapularis* and *Ixodes pacificus*. The pathogen cycles between wild animal hosts and vectors. Humans are accidental dead-end hosts. LD could involve the same agent *B. burgdorferi* but different vectors and hosts in different regions in the United States. In the northeastern United States, the white-footed mouse is the primary wildlife reservoir host responsible for infecting ticks (*I. scapularis*), also known as the deer tick (Jones et al. 1998).

Some infections that occur in northern California and the upper Pacific Northwest are transmitted by *I. pacificus* (Nielsen et al. 2008), the Western black-legged tick, and the main reservoir host is the dusky-footed wood rat. The main vector species in European countries are *Ixodes ricinus*. The main host of *I. ricinus* is the roe deer; although not a reservoir, it plays an important role of maintenance and cofeeding for ticks.

The process of LD extension beyond its endemic foci is predicted to accelerate with climate change. Research modeling climate change scenarios anticipate climate change will result in geographic distribution of vectors expanding northward as the earth warms, and they specifically forecast the retraction of vectors from the southern United States into the central United States and the emergence and reemergence of LD in various regions of Canada (Brownstein et al. 2004). Cases have been reported in over 60 countries, with endemic foci in North America, Europe, and Asia (WHO 2014).

Figure 4.4 shows that LD has extended to many countries around the world beyond the endemic foci (Ozdenerol 2015). Reported LD activities that were mapped include diagnosed cases as well as infected ticks, infected animals, and seropositive human samples. In response to the global expansion of LD, more studies are conducting spatial analysis to understand the conditions under which ticks spread, to highlight risk areas and determine environmental and climatic factors behind the prevalence of LD. Characteristics of these studies are summarized in Table 4.2.

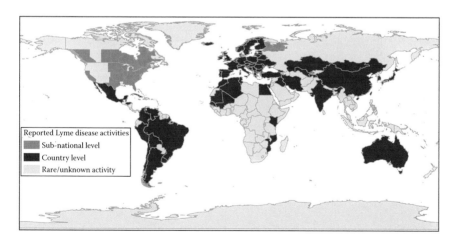

FIGURE 4.4
Geographic extension of Lyme disease (LD) activities. The figure shows that LD has extended to many countries around the world beyond the endemic foci. Reported LD activities that were mapped include diagnosed cases as well as infected ticks, infected animals, and seropositive human samples. The dark-gray shading signifies countries with (at least) some reported LD activity, and the presence of activity is known only at the country level. The lighter-gray shading represents areas in which LD has been reported at the subnational level in particular regions of some countries. The lightest gray represents counties with rare or unknown activity. (Courtesy of Ozdenerol, E., *Int. J. Environ. Res. Publ. Health*, 12(12), 15182, 2015.)

TABLE 4.2

Summary of Lyme Disease Studies with Common Risk Factors

GIS Analysis/Citation	Region/Date	Host, Vector, Pathogen	Data	Common Risk Factors
Geographic distribution				
GPS-based field data integrated into GIS (Pepin et al. 2012), Zonal statistics, Spatial autocorrelation, Predictive modeling, Density mapping	36 eastern states 2004–2006	*I. scapularis*	County level human case reports to the CDC as part of the national notifiable disease surveillance system (NNDSS) field-derived tick data Monthly vapor pressure, maximum daily temperature Normalized vegetation index (NASA) Elevation (USGS National land-cover database)	The distribution of *B. burgdorferi* genotypes. The estimated density of infected nymphs.
Risk mapping, (Guerra et al. 2002) GPS-based field data integrated into GIS, habitat analysis	Wisconsin, Illinois, Michigan 1996–1998	*I. scapularis* white-footed mice, chipmunks	Field-derived tick data Vertebrate collection Soil samples collection (top soil, leaf litter thickness, slope, pH, soil texture) Field-derived forest moisture index Climate data from NOAA (yearly and seasonal precipitation) Bedrock geology USDA Forest Service	Tick presence positively associated with deciduous, dry to mesic forests, and alfisol types of soils with loam-sand textures. Tick absence associated with grasslands, conifer forests, wet mesic forests, acidic soils of low fertility and a clay soil texture, Precambrian bedrock.
Geostatistics (Nicholson and Mather 1996) Spatial autocorrelation	Rhode Island	*I. scapularis*	State-wide collected human incidence data	A highly significant spatial trend for decreasing number of ticks and incident cases of LD with increasing latitude. Exposure to deer ticks and LD risk occurs mostly in the peridomestic environment.

(Continued)

TABLE 4.2 (*Continued*)

Summary of Lyme Disease Studies with Common Risk Factors

GIS Analysis/Citation	Region/Date	Host, Vector, Pathogen	Data	Common Risk Factors
Space–time scan statistics (Li et al. 2014)	Virginia 1998–2011	N/A	Census tract level count of LD human cases	Spatial expansion toward south and west along eastern coast of the United States areas where education and surveillance needs are the highest.
GPS-based field data integrated into GIS, density mapping (Ostfeld et al. 2006)	New York	Deer, mice, chipmunks	Growing season temperature, precipitation, abundance of hosts and acorns	Risk associated with prior year's abundance of mice and chipmunks and acorns.
Density surface mapping (Eisen et al. 2006) RS techniques Supervised classification	California Mendocino County	*I. pacificus*	Field-derived data: tree species, deer signs, NDVI, sunlight, hydrologic data	GIS-based environmental data could predict nymphal density more accurately than field-derived data.
Density surface mapping (Eisen et al. 2003) Habitat analysis	California Mendocino County	*I. pacificus*	Climatic variables, habitat type, deer usage on tick-related traits	A shift from peak nymphal densities occurring in oak woodlands in spring to redwood habitats in summer.
Clustering (Bunnell et al. 2003) Habitat analysis	Middle Atlantic region of United States 1997–1998	*I. scapularis*	Land cover, distance to water, forest edge, elevation and soil type	Clustered pattern along coastal plain of the Chesapeake Bay.
Spatial heterogeneity (Wimberly et al. 2008a,b; Estrada-Pena et al. 2013) Spatial autocorrelation Accuracy assessment			Species–habitat relationships Species–environment relationships	Spatial autocorrelation improves predictive spatial models.

(*Continued*)

TABLE 4.2 (*Continued*)

Summary of Lyme Disease Studies with Common Risk Factors

GIS Analysis/Citation	Region/Date	Host, Vector, Pathogen	Data	Common Risk Factors
Trends in surveillance				
Proximity analysis (Zeman and Benes 2014) Geographic stratification techniques	Czech republic 1997–2010	*I. ricinus*	Human population migration and demographic changes from Czech statistical office LD human cases	Population density, high incidence among 50–65-year-old people and 10-year-old children. Socioeconomic transformation. Amount of time people spent outdoors. Periresidential distribution and home vicinity.
Clustering analysis (Wilking and Strak 2014) Population density analysis	Germany 2009–2012	*I. ricinus*	Notified cases of LD clinical LD manifestations	Urban areas, forested areas, and public parks. Free-range livestock husbandry.
Clustering (Linard et al. 2007)	Belgium 1994–2004	*I. ricinus*	Deer population density Land use, forest cover	Human incidence, roe deer population, forest cover, population density, peri-urban areas.
Prevention behavior risk				
Surveys (Mckenna et al. 2004)	New York	N/A	Voluntary, anonymous questionnaire	Participants having a family member with LD were more likely to use preventive behaviors.
Cross-sectional (Schwartz and Goldstein 1990) Logistic regression	New Jersey 1988	*I. scapularis*	Occupation	Outdoor work.

(Continued)

TABLE 4.2 (*Continued*)

Summary of Lyme Disease Studies with Common Risk Factors

GIS Analysis/Citation	Region/Date	Host, Vector, Pathogen	Data	Common Risk Factors
Geographic stratification techniques (Bayles et al. 2013)	Missouri	*I. scapularis*	Structured interview park types	Human population density estimates.
Climate change				
A review of expert workshops, multivariate analyses, and predictions (Semenza and Menne 2009)	Belgium 2000–2010	*I. ricinus*	Incidence, prevalence, distribution of infections through various routes (vector, rodent, water, food, air)	Proposed to build an integrated network for environmental and epidemiologic data.
Wavelet-based time series analysis (Barrios et al. 2012) The NDWI. Cluster analysis of high disease risk Based on human cases			NDWI CORINE land-cover map obtained from European Environment Agency MODIS data (moderate resolution imaging spectroradiometer) obtained from Land Processes Distributed Active Archive Center	LD incidence. Vegetation greenness and moisture. Local characteristics of vegetative systems. Multiresolution analysis. Lagged climatic effects, vegetation, and moisture-related events spanning periods of 2 or more years.
Wavelet-based time series analysis (Barrios et al. 2013) The NDWI. Spatial autocorrelation Voronoi polygons Cluster analysis of high disease risk	Belgium 2003–2010	*I. ricinus*	Growing degree days (GDD) values calculated for each season derived from hourly temperature data from National Climatic Data Center and Royal Meteorological Institute of Belgium	Vegetated areas and frequent weather anomalies. GDD. Indicator of heat accumulation. Seasonal conditions affect the incidence.

(*Continued*)

TABLE 4.2 (Continued)

Summary of Lyme Disease Studies with Common Risk Factors

GIS Analysis/Citation	Region/Date	Host, Vector, Pathogen	Data	Common Risk Factors
Global climate modeling (Ogden et al. 2006) for two greenhouse gas emissions	Canada 1970–2000	*I. scapularis*	Grid point data of projected daily maximum and minimum temperatures obtained from two models: CGCM2 (Coupled Global Climate Modeling and Analysis) and UK Hadley Center's HadCM3 model	Annual degree days (DD > 0°C), seasonally variable temperature conditions. A2: Increasing heterogeneous population, fragmented economy, technology change. B2: Intermediate levels of economic growth and lower population growth.
Global climate modeling (IPCC 2000) for two greenhouse gas emissions	Canada 2020s, 2050s, 2080s	*I. scapularis*	Minimum temperatures obtained from two models: CGCM2 (Coupled Global Climate Modeling and Analysis)	Annual degree days (DD > 0°C), seasonally variable temperature conditions. A2: Range moved northwards by 200 km by the 2020s and 1000 km by the 2080s. B2: Projected expansion between 2050s and 2080s.
Risk mapping (Ogden et al. 2008) Simulation models. Validation through field studies	Canada 1970–2000 Projected 2020s, 2050s, 2080s	*I. scapularis*	Field-derived tick and rodent data Vertebrate collection Human Population data at census-subdivision level obtained from Statistics Canada Index numbers of ticks migrating on migratory birds Percentage cover of forest habitat	Vector populations, ambient temperature, number of nymphal ticks immigrating on migratory birds, and forest habitat cover. Predicted temperature conditions and emission scenarios.

(Continued)

TABLE 4.2 (Continued)

Summary of Lyme Disease Studies with Common Risk Factors

GIS Analysis/Citation	Region/Date	Host, Vector, Pathogen	Data	Common Risk Factors
Population genetic analysis				
GPS and field mapping (Ogden et al. 2006) Spatial expansion mapping (Kelly et al. 2014)	Virginia 2011	*I. scapularis*	Field-derived tick data Molecular methods	Population genetic signals of nymphal *I. scapularis*. Eastern most ticks with demographic expansion but not spatial expansion. Central and western tick populations with spatial expansion.
Host–pathogen relationship				
Range expansion mapping (Norris et al. 1996) Spatial structuring	European strains. Chinese strains	*Borrelia garinii Borrelia afzelii*	Multilocus sequence analysis Historical populations of *B. garinii* and *B. afzelii*	Geographic distances between collection sites. Rodent population expansions after the glacial maximum.
Prevalence mapping of antibodies (Vollmer et al. 2013) Clustering	Northeast, Upper Midwest. West Coast, United States	*B. burgdorferi*	County residence of each dog tested by zip code County level population data by U.S. census	Antibodies to *Borrelia* in dogs by zip code.
Finer-scale prevalence mapping (Bowman et al. 2009)	California	*B. burgdorferi*	CALVEG 2000 (vegetation coverage obtained from California Forestry and Fire protection) Precipitation isohyetal polygons	Seropositive and seronegative coyote locations, vegetation cover, and rainfall.
Vaccine deployment				
GPS and field mapping (Richer et al. 2014) Nymphal infection prevalence	New York	*B. burgdorferi*	Field-derived tick data	Significant decreases in tick infection prevalence were observed within 3 years of vaccine deployment.

Most of these studies attempt to quantify the associations between LD risk variables (e.g., vector, pathogen, and host abundance and distribution) and environmental variables using the spatial analysis capabilities of GIS. As ticks have limited mobility, their reservoir host range is an important spatial determinant. Long-term field monitoring of the host-seeking ticks in the host range helps define landscape predictors of LD risk.

In places where field data are not available, tick distribution is predicted based on *a priori* information about tick species ecology. RS and GIS are powerful tools for enabling the prediction of LD risk. Ticks' dependence on certain environmental factors (i.e., climatic factors) is experimentally verified through *a priori* approaches. Predictive maps of tick distributions are also produced by *ad hoc* statistical models (e.g., regression, discriminant analysis), based on GIS data reflecting the relationship between occurrence and a number of spatial covariates (i.e., vegetation, climatic, geological, and hydrological, soil types, host population covariates).

The breadth of GIS applications and approaches applying spatial analysis techniques to LD epidemiology varies. One interesting finding is that GIS-based environmental data could predict nymphal density more accurately than field-derived data (Eisen et al. 2006). Some studies are follow-up field studies integrating GPS-based field data into GIS (Guerra et al. 2002; Ostfeld et al. 2006; Pepin et al. 2012; Kelly et al. 2014; Richer et al. 2014).

A variety of data have been mapped, such as tick densities, pathogen genotypes, human incidence and population demographics, and host and vector habitats. A variety of mapping approaches have been applied such as density surface (tick) mapping (Eisen et al. 2003, 2006; Ostfeld et al. 2006; Pepin et al. 2012); risk mapping (Guerra et al. 2002; Ogden et al. 2008); population (human) density mapping (Wilking and Stark 2014); spatial expansion mapping of population (pathogen) genetic signals, range mapping of rodent populations, and prevalence mapping of antibodies (Norris et al. 1996; Foley et al. 2005); and prevalence of nymphal infection (Richer et al. 2014). Some studies use GIS to analyze and model habitats. GPS-based field data are integrated into a GIS for habitat analysis. They use land-cover data, proximity measures (distance to water, forest edge), soil variables (soil types, soil texture, soil pH), leaf litter thickness, slope and elevation data, and forest cover (tree species, forest types). Density surface mapping, zonal statistics, predictive modeling, and geostatistical tools are commonly used to perform spatial analysis (Guerra et al. 2002; Bunnell et al. 2003; Eisen et al. 2003; Nicholas et al. 2009).

Spatial variability in land cover, soils, and geology affect habitat suitability for vector species. Predictive modeling studies of tick and host distributions have applied spatial interpolation, spatial modeling, and spatial clustering techniques based on environmental indicators. Spatial autocorrelation improved predictive spatial models (Wimberly et al. 2008a,b; Estrada-Peña et al. 2013). Risk maps are developed using information from vegetation cover, moisture conditions, and temperature variables derived from satellite imagery (Estrada-Peña 1997; Ostfeld et al. 2006).

To model and project the related potential alterations in climate change on the distribution of ticks, data sources from NASA, USGS, CDC, and NOAA (climatic and environmental data) have been utilized in the development of spatial models (Guerra et al. 2002; Pepin et al. 2012). Monitoring tick abundance provides epidemiologically relevant information as well as tick absence so that statistical methods and predictive models can reveal the relationships of biological distributions with climate covariates (Nicholson and Mather 1996; Guerra et al. 2002). Maximum daily temperature, growing season temperature, annual degree days (DD > 0°C), growing degree days, monthly vapor pressure (humidity measure), and yearly and seasonal precipitation were the variables used to model and project the related potential alterations in climate change on the distribution of ticks (Ogden et al. 2006; Ostfeld et al. 2006). In order to test the hypotheses of the connections between climate and LD, wavelet-based time series analysis, and spectral indices, such as the normalized difference vegetation index (NDVI) and normalized difference water index (NDWI) are routinely used as remotely sensed measures of vegetation greenness and moisture (Eisen et al. 2003; Barrios et al. 2012). Spatial interpolation of climate data has been investigated as complementary approaches to predict spatial variations in frequent weather anomalies such as monthly climate (Ostfeld et al. 2006; Pepin et al. 2012; Barrios et al. 2013).

There are a number of potential issues and limitations of spatial analysis on LD. Cases are reported on the basis of the patient's residence rather than on the location in which the exposure occurred. Therefore, LD in a traveler returning from an area in which the disease is highly endemic cannot be construed as evidence of local transmission. Spatiotemporal component could provide misleading results because of the movement of the population between the time of infection and the onset of symptoms (Foody 2006).

Inconsistent methods of tracking human cases (e.g., as case numbers rather than incidence or incidence rates) and incomplete disease reporting of confirmed cases could result in fluctuations in case counts and reported rates, which in many instances vary between provinces or states within a country. Overreporting in nonendemic areas and underreporting in endemic areas could cause spatially biased results (Bacon et al. 2008).

More standardized data collection and analysis methods are needed given the current limitations of data collection and inconsistent tracking methods. National, provincial, and municipal boundaries are used for counts of human cases as part of notifiable disease surveillance systems (NNDSS) for mapping human incidence data. These administrative boundaries are arbitrary boundaries that do not coincide with biologic boundaries. This presents a potential problem since biologic boundaries contain ecological conditions of habitats, which affect the distribution and density of vectors and host animals involved in the transmission.

One of the drawbacks of researching LD epidemiology is that there are still not good baseline data sets available on vector surveillance. Ground-verifying field studies are time consuming and expensive. Empirical data on tick density and tick infection rates are difficult to collect for large areas. Therefore, there are very few large-scale studies researching the geographical variation in the relationship between human case surveillance data and tick densities (Pepin et al. 2012).

Region-specific standardized data collection and analysis approaches are needed to identify the determinants of spatial variation in LD risk and incidence. For example, sociocultural factors, recreation activities, demographics, and urbanism patterns influence humans and, in turn, occurrence of LD cases. The times of highest entomologic risk determine when best to do vector surveillance. However, factors affecting the entomological risk are location dependent. Local weather patterns (e.g., temperature, humidity) are influential for tick survival. Potential reservoir hosts for the North American LD system are rodents (e.g., white-footed mice, eastern chipmunks) and their abundance are strongly affected by their habitat variables including abundant food, forest cover, and nesting site conditions. These factors affect the reservoir host–tick–human interaction and consequently cause a great deal of variation in the distribution of LD within endemic zones.

For future implications, efforts to model climate change with more sophisticated RS data will improve our understanding of LD transmission cycles, identify risk areas, and assess their characteristics. The use of satellite imagery, GIS, and spatial statistical methods in conjunction with ground-verifying ecologic studies and LD case surveillance data has provided promising developments in LD research. Since LD transmission depends on complex ecological systems involving more than one agent, vector, and host and since it is regionally variable, spatial analysis of LD should adopt complementary approaches to geography, GIS, and spatial epidemiology from other disciplines including ecology, entomology, zoology, climatology, and virology.

Future research should also focus on long-term data collection that provides wide coverage of environmental and climatic, biotic and abiotic variables, and consequently contribute to the progress made in identifying the determinants of spatial variation in LD risk. The geographical and annual variation in the timing of human LD can be largely explained by weather conditions. Many GIS- and RS-based spatial models are developed for this. As availability of seasonal (temporal) and long-term spatial data increases, the quality and accuracy of GIS and RS methods improve, so does the effectiveness of spatial analysis.

The significant variability in seasonal as well as spatial risk of exposure to LD within small, but ecologically diverse, geographic areas shows that temporally dynamic and spatially explicit models are needed to assess the risk of exposure to tick-borne pathogens at spatial scales encompassing diverse climatologic or ecological conditions. Scale matters in predictive models. Variables with small spatial variance (i.e., macroclimatic conditions) in small areas have nearly no predictive value, whereas diversified variables (i.e., vegetation type) have limited value in large-scale studies. This aspect highlights the significance of multiresolution analysis and long-term monitoring studies for possible lagged climatic effects on the geographic distribution of vector populations. We need high-resolution (fine scale such as zip codes) human incidence data to help reveal isolated endemic areas. New GIS- and RS-based studies are needed to monitor occurrence at the macrolevel, and GPS-based field studies help pinpoint areas of occurrence at the microlevel where spread within populations of reservoir hosts, clusters of infected ticks, and tick-to-human transmission may be better understood.

Modeling risk based on habitat alone without follow-up data on the distribution of vectors and human cases would not be valid without testing the adequacy of these variables. Therefore, robust field studies are needed to validate and refine the risk maps. These risk maps can lead to identifying new endemic areas. Surveillance methods could be targeted for tick vectors in these expanded regions. Spatial models that use coarse scales and general climate and vegetation indexes fail to capture the complex relationship between tick activity and its field environment (Schulze et al. 2009).

Ground-verifying ecologic studies and acarological follow-up studies are crucial for effective control measures. Laboratory studies and spatial entomological and ecological risk models might show clear relationships between climatic variables (i.e., relative humidity) and tick survival but follow-up field studies might produce conflicting results. Host abundance patterns might have not been accounted for within field studies and could provide a limitation for verification of findings (Berger et al. 2014).

For future applications, more efficient control measures could be implemented with the aid of research outcomes from spatial analysis. GIS and spatial analysis could take a role in the optimal distribution strategies of the vaccines, such as locating bait stations containing a pesticide delivery system. To reduce vector-tick populations and human–tick encounters, by means of host-targeted methods, new GIS-based suitability models should be developed for effective vaccine deployment. Future developments will further enhance the novel use of GIS, merging data from various sources into an end-product tailored specifically to vaccine deployment.

Spatially comprehensive studies are needed for strategic implementation of intervention in LD endemic areas. An important conclusion pertaining to human populations considers that people living in areas where LD was not thought to be endemic may also be at risk for infection. New prevalence studies should be conducted on newly identified areas of endemicity.

Policies directing public health objectives in minimizing risk from LD could also include partnering with the tourism sector in disease surveillance by monitoring and reporting field conditions at high-risk recreation areas.

Waterborne Diseases

During the past few decades, human development, population growth, extreme weather events, natural calamities, and climate change have exerted many diverse pressures on both the quality and quantity of water resources that may in turn impact conditions fostering waterborne diseases. Access to adequate improved water supplies in informal settlements in many developing countries has important health dimensions (Crow and Odaba 2010). Worldwide, waterborne infectious diseases are a major cause of morbidity and mortality (Murray and Lopez 1997; Fenwick 2006; Lewin et al. 2007).

Waterborne infectious diseases are caused by pathogenic microbes that can be directly spread through contaminated water. Most waterborne diseases cause diarrheal illness. Cryptosporidiosis, diarrhea, cholera, and giardiasis are examples of waterborne diseases. Waterborne infectious diseases are acquired by drinking water contaminated at its source or in the water distribution system or by direct contact with environmental and recreational waters. Sources of contamination of drinking water include humans, domestic, and wild animals (Odoi et al. 2004). Water pollution results in human health risks by exposure not only through drinking water from surface water or groundwater supplies but also through organisms in the food chain that may be contaminated with chemical carcinogens.

GIS is used at a macroscale for global and regional mapping of "hot spots" of water supply in relation to population growth, human development, and changing climate (Yang et al. 2012). A raster GIS approach using satellite imagery and band algebra (e.g., NDVI) techniques is employed to differentiate between land and water, identify surface water features (e.g., rivers, canals, ponds, tanks, swamps), create distance surfaces, and measure water quality characteristics (Ali et al. 2002). A GIS overlay analysis was employed to generate a map of the relationship between the availability of terrestrial surface water and distribution of water-associated diseases at the global scale (Figure 4.5). The primary source of information on water-associated pathogens and infectious diseases in this map was based on the Global Infectious Disease and Epidemiology Network, a web-based comprehensive global infectious diseases database (Smith et al. 2007). Each disease was classified into one of the following five categories: waterborne, water based, water related, water washed, and water dispersed. Waterborne diseases, such as typhoid and cholera, are typically caused by enteric microorganisms, which enter water sources through fecal contamination and cause infections in humans through ingestion of contaminated water. To account for waterborne pathogens (e.g., *Cryptosporidium*, *Giardia*) whose transmission can be through accidental ingestion of, or exposure to, contaminated water in recreational settings, those outbreaks caused by this transmission pathway are included in "water-carried diseases"; water-based diseases commonly

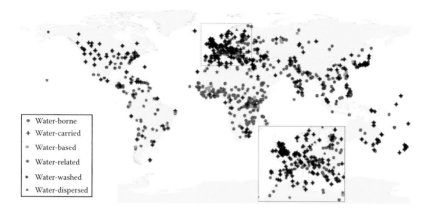

FIGURE 4.5
Distribution of reported outbreaks of water-associated infectious diseases from 1991 to 2008. (Courtesy of Yang, K. et al., *PLoS Negl. Trop. Dis.*, 6(2), e1483, 2012.)

refer to diseases caused by infections of worms that must spend parts of their life cycles in the aquatic environment, such as schistosomiasis; water-related diseases, such as malaria and trypanosomiasis, need water for breeding of insect vectors to fulfill the transmission cycle; water-washed diseases are those whose transmission is due to poor personal and/or domestic hygiene as a result of lack of appropriate water; and finally, water-dispersed diseases are caused by infections of agents that proliferate in freshwater and enter the human body through the respiratory tract, such as *Legionella*. GIS and spatial analytical techniques offer means for developing and organizing spatially explicit information to understand these transmission pathways, the spatial structure associated with the distribution of the outbreaks, and underlying risk factors (McMichael 2004).

Waterborne disease is less likely to produce a clear pattern of diffusion over space and time. Studying such space–time patterns requires epidemiological databases and location information, where the cases or deaths were recorded as well as when they were observed (Lawson and Kleinman 2005).

John Snow identified the outbreak of cholera by mapping the affected London water sources in 1854. He was able to correlate the case locations to a centrally located water pump and confirmed the contaminated water was responsible for the spread of the disease. Studying the spatial distribution of cases can be useful in identifying geographical variation, but not necessarily for identifying the reasons for this variation. Population at risk and the progression of the epidemic over time should also be investigated. In a recent study, Shiode et al. (2015) revisited Snow's study and examined the space–time pattern of the cholera outbreak in GIS using historical documents. Mortality rates and the space–time pattern in the victims' records were explored using kernel density estimation (KDE) and network-based scan statistic (NetScan), a recently developed method that detects significant concentrations of records such as the date and place of victims with respect to their distance from others along the street network. The results are visualized in a map form using a GIS platform. Figure 4.6 shows an overlay of KDE- and the NetScan-detected clusters confirming high mortality rates identified around the Broad Street pump.

GIS aids the identification of possible associations between disease and particular water supplies (Hughes et al. 2004). GIS is essential for running a waterborne disease surveillance system. A GIS-based surveillance system can only be as good as the collected databases. The data sets should comprise not only epidemiological databases but also information about water supply, water treatment, and distribution. In mapping and analyzing waterborne diseases, the water distribution network, including reservoirs, mains, and public and private wells, is crucial. The input of demographic

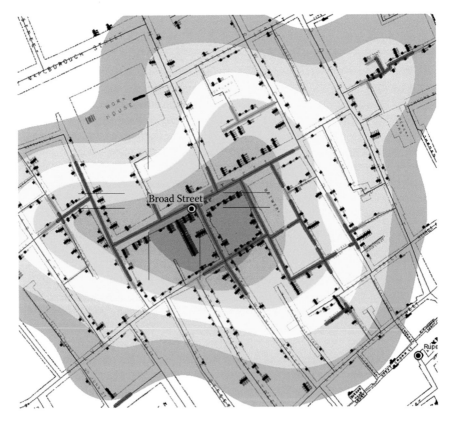

FIGURE 4.6
An overlay of kernel density estimation and the clusters detected through NetScan. (Courtesy of Shiode, N. et al., *Int. J. Health Geogr.*, 14, 21, 2015.)

and epidemiological data into a GIS linked with water quality modeling provides new insights in waterborne disease patterns. GIS techniques support the generation of hypotheses regarding waterborne disease distribution and causation. They are extremely useful to carry out area-based correlation studies, point source pollution studies, and analyze the exposure of populations. Parameters depicting the water supply structures, especially the amount of drinking water produced from surface water or groundwater and the amount of people supplied by the surface water or groundwater, were correlated with the age-standardized incidence rates of gastrointestinal infections (Dangendorf et al. 2002). Other parameters, which characterize distribution networks in detail such as pipe length, age, dead ends, material, and other distribution facilities, are included in modern GIS database systems and are part of the daily workflow for many utilities and public works systems. However, administrative unit-level analysis makes it hard to include these detailed water supply structures and related factors to the residential populations (Havelaar 1994). For highest likelihood of analysis success, the correlation models should be carried out on the smallest available units such as subdivision or block level.

GIS is also used at micro (neighborhood/subdivision)-scales for water distribution systems, mapping and quantifying of water availability (the absolute amount that can be sustainably appropriated from surface water or groundwater stores), delivery systems (infrastructure), and quality measurements (the treatment and suitability of water for domestic use). The continued monitoring of all facilities for issues such as their cost, quality, functionality, and sustainability becomes a huge task and can be handled efficiently by employing GIS. The public drinking water system databases could be acquired through state engineering services and/or safe drinking water programs in the United States. The database includes water wells, their latitude and longitude, sampling data inorganics and corrosivity, testing for volatile organics, and PCBs. The wells can be mapped through each well's locational coordinate in a GIS. A relational join of the database tables is created to link the attribute data to the point wells. These data then integrated with epidemiologic and demographic databases.

The results of water quality analyses can be checked by regional comparisons and temporal trends through GIS mapping.

Airborne Diseases

Airborne transmission occurs when pathogens spread from host to host through the process of respiration, as occurs with influenza, tuberculosis, or the common cold. Airborne diseases emerge out of geographical patterns of human contact and interaction, which are linked to poor-quality housing conditions, crowding, and personal contact in facilities such as schools, day-care centers, jails, homeless shelters, and workplaces. Population density, poverty, and overcrowding appear in most areas to be major factors for airborne disease transmission (Acevedo-Garcia 2001). For example, urban centers have traditionally higher rates of tuberculosis than rural areas (Barnes and Barrows 1993). Identifying persons with recently acquired infections is an important component of outbreak investigations and strategic TB elimination. GIS-based screening and treatment could be an effective method for TB control programs to identify high-risk populations. In addition to traditional surveillance, identifying specific strains through the use of molecular strain characterization methods yields insight into TB transmission. Moreover, examining how molecular clustering varies spatially by mapping the genotypically clustered cases to the total number of cases reported could demonstrate that molecularly clustered disease is not homogeneously distributed. The majority of the tuberculosis morbidity within the developed world is strongly influenced by imported tuberculosis from high prevalence countries, though the rates of transmission to the general population remain low (Zuber et al. 1997). Foreign-born cases are significantly more likely to have a unique strain (Moonan et al. 2004). Using GIS, identifying geographic areas of increased incidence with a high percentage of unique strains may improve local surveillance methods to locate hard-to-reach foreign-born populations before transmission occurs (Moonan et al. 2004).

Air quality, solar radiation, radon gas levels, and exposure to outdoor air pollutants, potentially harmful to human health can be monitored and brought into a GIS and integrated analytically with other health predictor variables and outcome data. Exposure to outdoor air pollutants, including ozone (O_3), particulate matter, and hazardous air pollutants (HAPs) are known risk factors for developing respiratory diseases such as asthma (Strachan 2000). Older and substandard housing are more likely to contain known triggers of asthma, such as mold and mildew, cockroach and rodent allergens, and dust mites.

Using GIS, we investigate the role of neighborhood environment on elevated childhood asthma hospitalization rates such as neighborhood-specific hazards, including housing and environmental exposures that may contribute to the onset and triggering of the disease. State health departments in the United States consolidate hospitalization data by aggregating discharge diagnosis records to upper levels of administrative units (i.e., census tracts) and calculate hospitalization rates. GIS is used to incorporate potential air pollution contributions from multiple sources. Since actual ambient air-monitoring data at the neighborhood level do not exist, proxy variables are included in the analysis such as density of stationary source air polluting facilities, polluting land uses, and truck routes for estimating mobile source air pollution.

The two environmental databases deemed most appropriate for estimating the respiratory burden from stationary source air pollution in the

United States are (1) the U.S. Environmental Protection Agency (EPA), Aerometric Information Retrieval System (AIRS) Facility Subsystem (http:// www.epa.gov/enviro/html/airs/airs_quaery_java.html) (these data include small stationary air polluting facilities such as boilers in apartment buildings and dry cleaners), and (2) the U.S. EPA, Toxic Release Inventory system (http://www.epa.gov/tri), specifically, stack and fugitive emissions (EPA 2015a). These are large facilities that emit larger amounts of toxic pollutants into the air than those in the AIRS database. The U.S. EPA produces a comprehensive National Emissions Inventory (NEI) for criteria air pollutant and HAP every 3 years from all air emissions sources (please see http:// www3.epa.gov/ttnchie1/net/2011inventory.html) (NEI 2011). Facility locations expressed in latitude and longitude and the amount of the emission are provided for the major pollutants (e.g., CO [carbon monoxide], lead). This database is used to map air polluting facilities by pollution type. GIS are used to visualize emitting facility distribution and air pollution. Figure 4.7 (Anzhelika Antipova, personal communication, 2015). The smallest possible cell size and facility of interest are the most useful units of analysis to communicate a realistic measure of the potential for adverse impacts stemming

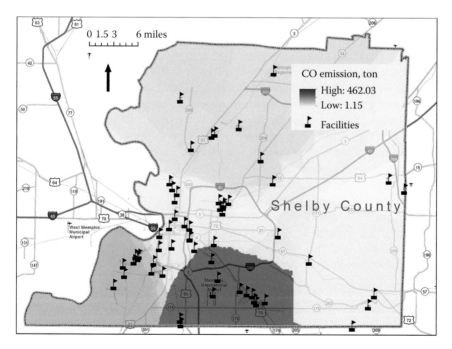

FIGURE 4.7
The distribution of air polluting facilities and a spatial interpolation of CO emission in Shelby County, Tennessee, for year 2008. Areas with high values of CO emissions are in darker color. There were 61 facilities reported for Shelby County in year 2008. (Courtesy of Anzhelika Antipova).

from exposure to these emissions. In order to accurately measure a region's vulnerability, release quantities with measures of toxicity, estimates of the number of people potentially exposed, and the contributions of nearby areas to the local vulnerability could be consolidated in an integrated GIS system.

The age and housing quality data could be retrieved from City Housing and Vacancy Surveys, such as in New York City Region (Census 2000). Public Housing Units data, including a crowding variable, are collected by the U.S. Department of Housing and Urban Development. These data sets could be obtained via the Census Data (HUD 2015) (http://www.census.gov/hhes/www/housing.html). Noxious land uses, zoned for manufacturing and industry that contain gas stations, power plants, sewage treatment plants, bus depots, rail yards, and vacant properties, are considered polluting land uses. The density of the number of potentially polluting land uses per square mile for each census tract is calculated in a GIS as a composite variable.

For socioeconomic data, the smallest practicable unit of analysis yields the most accurate and realistic results (Krieger et al. 2002, 2003). Urban asthma is a growing epidemic among African-American and Latino children in the United States. African-American children living below the poverty line were 15%–20% more likely to have asthma (CDC 2007). Subsequent asthma surveillance reports confirmed these differences and documented that the differences have persisted over time (Evans et al. 1987; Mannino et al. 2002; CDC 2007). These reports indicate that population-based asthma prevalence rates, emergency department visit rates, and hospitalization rates were higher among blacks than among whites, higher among females than among males, higher among children than among adults, and higher among males aged 0–17 years than among females in the same age group. In addition, more detailed analysis of ethnicity data demonstrated that different Hispanic groups had differing health outcomes. Among Hispanics, those of Puerto Rican descent (origin or ancestry) had higher asthma prevalence and death rates than other Hispanics (e.g., those of Central and South American descent), non-Hispanic blacks, and non-Hispanic whites (Homa et al. 2000; Rose et al. 2007). Besides race and ethnicity, other SES variables found to be associated with asthma rates include median household income, the percentage of the neighborhood population in poverty, and the mean household rent.

There are inherent limitations reflected in the reality faced in GIS analysis of urban asthma investigations. Asthma hospitalization data are used as a surrogate for asthma prevalence, which do not include individual patient data and come as the number of admissions, sometimes not able to screen out multiple admissions of the same asthmatic patient. Individual-patient-record-level data are the most useful unit of analysis for health data. However, because of patient confidentiality, individual asthma hospitalization cases are aggregated to the nearest unit of analysis (e.g., tracts and block groups) they fell in. Positional errors of the point data (where points could be assigned to the incorrect unit) affect the rates.

Estimating the aerial extent of exposure, the unit of spatial data aggregation, spatial autocorrelation, level of resolution, and delineation of the optimal study area extent are geographical considerations researchers encounter while performing analysis of neighborhood effects on asthma (Corburn et al. 2006). Spatial data by its nature are not randomly distributed, while traditional statistical approaches require randomness. Therefore, spatial autocorrelation becomes an impediment to the application of traditional statistical approaches (Tobler 1979). The principle of spatial autocorrelation states that "Features that are similar in location also tend to be similar in attributes." Accounting for spatial autocorrelation is one of the relatively small sets of techniques that deals simultaneously with both locational and attribute information (Goodchild 1986). The added benefit of accounting for spatial autocorrelation in analyses needs to be investigated.

Different units of analysis yield different results, and results are also influenced by the method of assessing exposure potential (McMaster et al. 1997). Differences in how the units are combined can produce substantial differences in the portrayal of the prevalence in minority populations relative to the locations of pollution sites. This has implications for environmental justice issues, where particular social groups have disproportionate burdens from environmental hazards (Pellow 2000). For example, populations within a U.S. zip code containing a pollution site could be considered as impacted. One who is within the same zip code but lives far from the pollution site is still considered to be impacted. Populations within a buffer distance are considered impacted though it assumes everyone within the buffer is impacted equally, when we know air pollution does not disperse equally in all directions from a source (Maantay 2007). A plume buffer based on results from an air dispersion model estimates the extent and direction of pollutant dispersion as well as pollutant concentration levels. With the buffer analysis (circular buffer), the impacted population may include a low percentage of minorities. However, with a more precise air dispersion model showing actual contaminant concentration contours, it may be the case that the area of highest impact from a facility's pollutant emissions is an area occupied by a high percentage of minorities.

Sexually Transmitted Diseases

STDs are illnesses that are spread by sexual contact. Understanding the transmission pathways of infection is critical in any GIS assessment of STDs. Intimate relations that channel STD spread are not clearly mappable as traditional spatial patterns of human interaction. The implications for mapping depend on the geographical distribution of reporting bias. The figures for the prevalence and incidence of even the monitored STDs are inexact because of asymptomatic or unreported STDs. The prevalence of STDs is much higher than their incidence. Economic deprivation at the individual, household, and community level increases the likelihood of contracting an STD and having it reported to a public health agency. Significant underreporting exists for STDs. STDs also show strong class differences in underreporting. High-income people are more likely to visit private physicians than public clinics. This class-based disparity leads to lower rates of reporting in high-income areas and exaggerates the apparent degree of clustering in low-income areas.

STDs tend to coexist. Behavior that puts a person at risk for one STD may predispose them to having more than one infection (CDC 2012). GIS is applied to examine whether each reported STD has a clustered spatial distribution and whether core areas of infection overlap for various STDs (Zenilman et al. 1999). Figure 4.8 shows an example map of coinfection rate. Nonuniform distribution of STD prevalence suggests that STD-specific prevention strategies should be targeted in core areas. Overlap of core areas among STDs suggests that intervention and prevention strategies can be combined to target multiple STDs effectively.

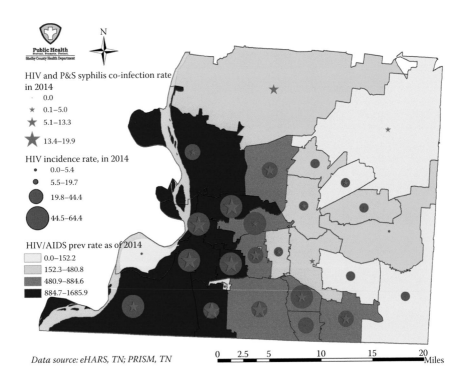

FIGURE 4.8
HIV and primary and secondary (P & S) syphilis coinfection rate in 2014. (Courtesy of Memphis and Shelby County Health Department, Memphis, TN.)

STDs are widespread and they are among the most common infectious diseases in the world. In some developing countries, HIV/AIDS has become the leading cause of death (WHO 2012a). Currently, STDs occur more frequently in certain identifiable subpopulations. The differences in behavioral factors between subpopulations reflect larger differences in STD prevalence. In the United States, there are higher rates of STDs among African-American and Hispanic males. Sixty percent of women with AIDS are African-American (Dariotis et al. 2011; Mojala and Everett 2012; WHO 2012b,c). Sub-Saharan Africa has by far the largest HIV epidemic in the world, with 68% of the global HIV disease burden and 1.9 million new infections in 2010 (UNAIDS 2010).

GIS has been widely used in analyzing the spatial distribution of HIV infection, clarifying specific drivers of HIV transmission and identifying priority populations for HIV prevention interventions, and planning of HIV services. GIS could provide more accurate information about HIV/AIDS incidence and prevalence rates and delineate maps of high disease-burden areas. Smallman et al. (1992) in their atlas have mapped AIDS and representation of disease events. Jones et al. (2003) added temporal component to the AIDS maps. Casetti and Fan (1991) developed spatiotemporal models for AIDS cases in the United States. Nagelkerke et al. (2002) developed dynamic simulation models for the country of Botswana as a prevention strategy. Nakaya et al. (2005) applied spatiotemporal modeling of AIDS in Japan using spatial parameters. GIS is used to map the number of HIV services in high disease-burden areas to understand access to prevention and health care (Law et al. 2004). A GIS-based spatial information management system containing infrastructure and AIDS patient data plays a vital role in the decision-making of governments on determining when and where to intervene, improving the quality of care for HIV patients, increasing accessibility of services, and delivering a cost-effective mode of information. However, African governments face significant barriers to the implementation of GIS, due both to a lack of STD data (because of limitations in disease detection and reporting systems) and to the nonexistence of detailed digital maps, especially in areas affected by conflict, population displacement, and rapid urbanization. However, through spatially targeted preventions, behavioral or biomedical interventions in high disease-burden areas may have considerable impact in reducing the sizes of the HIV-AIDS epidemic in Africa and the developed world (Cuadros et al. 2013).

Drug Addiction

GIS analysis and spatial analytical models are relevant geographic techniques to better understand the epidemiology of drug addiction. GIS is applied to assess the links between social interactions, physical distance, and drug using behavior. Geographic and neighborhood context variables that have been studied in this regard include distance from residence to outpatient treatment facility (Friedmann et al. 2001; Beardsley et al. 2003), distance from residence to risky or safe places (Mason et al. 2004), and proximity to local drug markets (Hunt and Kennedy). GIS allows computation of Euclidean distance between geo-referenced points (e.g., centroid of a neighborhood boundary or census tract), measure street network travel distances, and establish service area buffers and travel time contours around locations of interest, such as a treatment facility. Although there are many limitations in trying to obtain accurate positional data in drug use studies, applying GIS will substantially improve understanding of interactions between drug users and the at-risk environment. Using handheld GPS devices, the point locations of drug treatment programs, health centers, and intravenous drug users' residences and injection neighborhoods are collected. Attributes associated with each location are entered into a GIS, and then for each documented hospital case, evaluations can then take place of whether proximity to treatment centers is related to participation in a relevant program.

Although linkages between neighborhood environment and drug-using behaviors have been discussed conceptually (Tucker et al. 1991; Joe et al. 1994; Jacobson 2004), relatively few studies have examined this relationship empirically (Latkin et al. 1998; Mason et al. 2004). These studies have looked at related associations, for instance, the associations between both the percentage of alcohol users and the number of bars in a neighborhood and the number of traffic injuries (Van Oers and Garretsen 1993); the association between alcohol outlet density and rates of violent assault (Scribner et al. 1995; Gruenewald et al. 2006); the association between neighborhood disadvantage and substance abuse rates and/or the increase in likelihood of mental health problems (Boardman et al. 2001); the effect of drug availability in neighborhoods on drug prices and substance abuse (Caulkins 1995); the effects on substance abuse of persons living in areas with high housing vacancies or physically deteriorated neighborhoods (Krause 1996); associations between alcohol availability, drug "hot spots," and violent crime (Zhu et al. 2006); examining patterns of illicit drug use (Dembo et al. 1985; Smart et al. 1994); and the association between school and neighborhood characteristics and school rates of substance use (Ennett et al. 1997). In these studies, GIS provides a framework for storing geographically referenced information on social contextual factors such as neighborhood characteristics, as well as a method of linking contextual factors with individual-level data (Harries 1995). Since individuals are nested within their local environment (e.g., neighborhood), and this nested structure implies that the observations based on individuals sharing the same local environment are not statistically independent (Williams et al. 2006), if one analyzes such data with conventional statistical methods that assume independent observations, results may be obscured or biased (Bryk and Raudenbush 1992). To avoid the nested structure, individual risk factors are frequently aggregated to a census geography (e.g., census tracts, block groups), and then the census tracts become the unit of observations, which are often considered statistically independent. The use of aggregated census-based variables in conventional multivariate analysis is subject to bias (Geronimus and Bound 1998; Davey-Smith and Hart 1999; Krieger 1999) and can lead to the ecological fallacy. However, new techniques in multilevel analysis allow one to distinguish between the relative contributions of within- and between-neighborhood effects as well as estimate how much variability at the contextual level is due to individual factors (Diez Roux 2001, 2004; Oakes 2004).

Conclusion

Several conclusions emerged from this chapter. First of all, understanding infectious disease starts with identifying its spatial characteristics. Spatial analysis of infectious diseases and vector-borne diseases contributes to epidemiologic knowledge of exposure to infection, which informs diagnostic testing and assists clinicians in the accurate diagnosis. Spatial analysis provides high levels of insight into understanding the conditions under which vectors spread, risk areas could be highlighted, and environmental and climatic factors behind the prevalence could be determined.

There are a number of potential issues and limitations of spatial analysis on infectious disease epidemiology. Cases are reported on the basis of the patient's residence rather than on the location in which the exposure occurred. Therefore, infectious disease, such as LD, in a traveler returning from an area in which the disease is highly endemic cannot be construed as evidence of local transmission. Spatiotemporal component could provide misleading results because of the movement of the population between the time of infection and the onset of symptoms (Foody 2006). Inconsistent methods of tracking human cases (e.g., as case numbers rather than incidence or incidence rates) and incomplete disease reporting of confirmed cases could result in fluctuations in case counts and reported rates, which in many instances vary between provinces or states within a country. Overreporting in nonendemic areas and underreporting in endemic areas could cause spatially biased results (Bacon et al. 2008). More standardized data collection and analysis methods are needed given the current limitations of data collection and inconsistent tracking methods. National, provincial, and municipal boundaries are used for counts of human cases as part of NNDSS for mapping human incidence data. These administrative boundaries are arbitrary boundaries that do not coincide with biologic boundaries. This presents a potential problem since biologic boundaries contain ecological conditions of habitats, which affect the distribution and density of vectors and host animals involved in the transmission.

References

Acevedo-Garcia, D. (2001). Zip-code level risk factors for tuberculosis: Neighborhood environment and residential segregation in New Jersey, 1985–1992. *American Journal of Public Health* 91(5): 734–741.

Ali, M., Emch, M., Donnay, J.P., Yunus, M., and Sack, R.B. (2002). Identifying environmental risk factors for endemic cholera: A raster GIS approach. *Health and Place* 8: 201–210.

Anderson, J.F., Andreadis, T.G., Vossbrinck, C.R., Tirrell, S., Wakem, E.M., French, R.A., Garmendia, A.E., and Van Kruiningen, H.J. (1999). Isolation of West Nile virus from mosquitoes, crows, and a Cooper's hawk in Connecticut. *Science* 286: 2331–2333.

Bacon, R.M., Kugeler, K.J., and Mead, P.S. (2008). Surveillance for Lyme disease, United States, 1992–2006. http://www.cdc.gov/MMWR/PREVIEW/MMWRHTML/ss5710a1.htm. Accessed January 11, 2015.

Barnes, P.F. and Barrows, S.A. (1993). Tuberculosis in the 1990's. *Annals of Internal Medicine* 119: 400–410.

Barrios, J.M., Verstraeten, W.W., Maes, P., Aerts, J.M., Farifteh, J., and Coppin, P. (2013). Seasonal vegetation variables and their impact on the spatio-temporal patterns of nephropathia epidemica and Lyme borreliosis in Belgium. *Applied Geography* 45: 230–240.

Barrios, J.M., Verstraeten, W.W., Maes, P., Clement, J., Aerts, J.M., Farifteh, J., Lagrou, K., van Ranst, M., and Coppin, P. (2012). Remotely sensed vegetation moisture as explanatory variable of Lyme borreliosis incidence. *International Journal of Applied Earth Observation and Geoinformation* 18: 1–12.

Bayles, B.R., Evans, G., and Allan, B.F. (2013). Knowledge and prevention of tick-borne diseases vary across an urban-to-rural human land-use gradient. *Ticks and Tick-Borne Diseases* 4: 352–358.

Beardsley, B.A., Wish, E.D., Fitzelle, D.B., O'Grady, K., and Arria, A.M. (2003). Distance traveled to outpatient drug treatment and client retention. *Journal of Substance Abuse and Treatment* 25: 279–285.

Berger, K.A., Ginsberg, H.S., Gonzalez, L., and Mather, T.N. (2014). Relative humidity and activity patterns of *Ixodes scapularis* (Acari: Ixodidae). *Journal of Medical Entomology* 51(4): 769–776.

Bertoletti, L., Kitron, U., and Goldberg, T.L. (2007). Diversity and evolution of West Nile virus in Illinois and the Unites States, 2002–2005. *Virology* 360: 143–149.

Bertoletti, L., Kitron, U.D., Walker, E.D., Ruiz, M.O., Brawn, J.D., Loss, S.R., Hamer, G.L., and Goldberg, T.L. (2008). Fine-scale genetic variation and evolution of West Nile virus in a transmission "hot Spot" in suburban Chicago, USA. *Virology* 374: 381–389.

Bian, L. and Li, I. (2006). Combining global and local estimates for spatial distribution of mosquito larval habitats. *GIScience and Remote Sensing* 43: 95–108.

Boardman, J.D., Finch, B.K., Ellison, C.G., Williams, D.R., and Jackson, J.S. (2001). Neighborhood disadvantage, stress, and drug use among adults. *Journal of Health and Social Behavior* 42: 151–165.

Bouden, M., Moulin, B., and Gosselin, P. (2008). The geosimulation of West Nile virus propagation: A multi-agent and climate sensitive tool for risk management in public health. *International Journal of Health Geographics* 7: 35.

Bowman, D., Little, S.E., Lorentzen, L., Shields, J., Sullivan, M.P., and Carlin, E.P. (2009). Prevalence and geographic distribution of *Dirofilaria immitis, Borrelia burgdorferi, Ehrlichia canis*, and *Anaplasma phagocytophilum* in dogs in the United States: Results of a national clinic-based serologic survey. *Veterinary Parasitology* 160: 138–148.

Brown, H.E., Childs, J.E., Diuk-Wasser, M.A., and Fish, D. (2008). Ecologic factors associated with West Nile virus transmission, northeastern United States. *Emerging Infectious Diseases* 14: 1539–1545.

Brownstein, J.S., Holford, T.R., and Fish, D. (2004). Enhancing West Nile virus surveillance, United States. *Emerging Infectious Diseases* 10: 1129–1133.

Bryk, A.S. and Raudenbush, S.W. (1992). *Hierarchical Linear Models: Applications and Data Analysis Methods*. Newbury Park, CA: Sage.

Bunnell, J.E., Price, S.D., Das, A., Shields, T.M., and Glass, G.E. (2003). Geographic information systems and spatial analysis of adult *Ixodes scapularis* (Acari: Ixodidae) in the Middle Atlantic Region of the U.S.A. *Journal of Medical Entomology* 40(4): 570–576.

Calistri, P., Ippoliti, C., Candeloro, L., Benjelloun, A., Harrak, M.E., Bouchra, B., Danzetta, M.L., Sabatino, D.D., and Conte, A. (2013). Analysis of climatic and environmental variables associated with the occurrence of West Nile virus in Morocco. *Preventive Veterinary Medicine* 110: 549–553.

Casetti, E. and Fan, C.C. (1991). The spatial spread of the AIDS epidemic in Ohio: Empirical analysis using the expansion methods. *Environment and Planning A* 23: 1589–1608.

Caulkins, J.P. (1995). Domestic geographic variation in illicit drug prices. *Journal of Urban Economics* 37: 38–56.

CDC. (2007). National surveillance for asthma—United States, 1980–2004. *Morbidity and Mortality Weekly Report* 56(SS-8): 1–14, 18–54.

CDC. (2012). STDs: Making the connection. http://www.cdcnpin.org. Accessed December 12, 2015.

CDC. (2013). CDC and USGS have employed GIS and RS to prepare interpretive maps showing WNV activity. http://www.cdc.gov/ncidod/dvbid/westnile/resources/wnvguidelines1999.pdf. Accessed September 23, 2013.

Census. (2000). New York City region. Housing & Vacancy Survey (Local Census). Washington, DC: Bureau of the Census, U.S. Department of Commerce.

Chuang, T., Henebry, G.M., Kimball, J.S., Van Roekel-Patton, D.L., Hildreth, M.B., and Wimberly, M.C. (2012). Satellite microwave remote sensing for environmental modeling of mosquito population dynamics. *Remote Sensing of Environment* 125: 147–156.

Cleckner, H.L., Allen, T.R., and Bellows, S. (2011). Remote sensing and modeling of mosquito abundance and habitats in Coastal Virginia, USA. *Remote Sensing* 3: 2663–2681.

Cooke, W.H., Grala, K., and Wallis, R.C. (2006). Avian GIS models signal human risk for West Nile virus in Mississippi. *International Journal of Health Geographics* 5(1): 36.

Corburn, J., Ozleeb, J., and Porter, M. (2006). Urban asthma and the neighborhood environment in New York City. *Health and Place* 12: 167–179.

Crow, B. and Odaba, E. (2010). Access to water in a Nairobi slum: Women's work and institutional learning. *Water International* 35(6): 733–747.

Cuadros, D.F., Awad, S.F., and Abu-Raddad, L.J. (2013). Mapping HIV clustering: A strategy for identifying populations at high risk of HIV infection in sub-Saharan Africa. *International Journal of Health Geographics* 12: 28.

Dangendorf, F., Herbst, S., Reintjes, R., and Kistemann, T. (2002). Spatial patterns of diarrheal illnesses with regard to water supply structures—GIS analysis. *International Journal of Hygiene and Environmental Health* 205: 183–191.

Dariotis, J.K., Sifakis, F., Pleck, J.H., Astone, N.M., and Sonenstein, F.L. (2011). Racial-ethnic disparities in sexual risk behaviors and STDs during the transition to adulthood for young men. *Perspectives on Sexual and Reproductive Health* 43(3): 53–59.

Davey-Smith, G. and Hart, C. (1999). Re: Use of census-based aggregate variables to proxy for socioeconomic group: Evidence from national samples. *American Journal of Epidemiology* 150: 996–997.

David, S.T., Mak, S., MacDougall, L., and Fyfe, M.A. (2007). Bird's eye view: Using geographic analysis to evaluate the representativeness of corvid indicators for West Nile virus surveillance. *International Journal of Health Geographics* 6: 3.

Davis, C.T., Ebel, G.D., Lanciotti, R.S. et al. (2005). Phylogenetic analysis of North American West Nile virus isolates, 2001–2004; Evidence for the emergence of a dominant genotype. *Virology* 342: 252–265.

De Groote, J.P., Sugumaran, R., Brend, S.M., Tucker, B.J., and Bartholomay, L.C. (2008). Landscape, demographic, entomological and climatic associations with human disease incidence of West Nile virus in the state of Iowa, USA. *International Journal of Health Geographics* 7: 19. doi:10.1186/1476-072X-7-19.

Dembo, R., Schmeidler, J., Burgos, W., and Taylor, R. (1985). Environmental setting and early drug involvement among inner-city junior high school youths. *International Journal of Addiction* 20(8): 1239–1255.

Diez Roux, A.V. (2001). Investigating neighborhood and area effects on health. *American Journal of Public Health* 91(11): 1783–1789.

Diez Roux, A.V. (2004). Estimating neighborhood health effects: The challenges of causal inference in a complex world. *Social Science and Medicine* 58(10): 1953–1960.

Eisen, R.J., Eisen, L., Castro, M.B., and Lane, R.S. (2003). Environmentally related variability in risk of exposure to Lyme disease spirochetes in Northern California: Effect of climatic conditions and habitat type. *Environmental Entomology* 32(5): 1010–1018.

Eisen, R.J., Eisen, L., and Lane, R.S. (2006). Predicting density of *Ixodes pacificus* nymphs in dense woodlands in Mendocino County, California, based on geographic information systems and remote sensing versus field-derived data. *American Journal of Tropical Medicine and Hygiene* 74(4): 632–640.

Ennett, S.T., Flewelling, R.L., Lindrooth, R.C., and Norton, E.C. (1997). School and neighborhood characteristics associated with school rates of alcohol, cigarette, and marijuana use. *Journal of Health and Social Behavior* 38(1): 55–71.

EPA. (2015a). Toxic release inventory (TRI) [Online]. http://www.epa.gov/tri. Accessed December 7, 2015.

EPA. (2015b). Aerometic information retrieval system (AIRS) facility subsystem [Online]. http://www.epa.gov/enviro/html/airs/airs_quaery_java.html. Accessed December 1, 2015.

Epp, T.Y., Waldner, C.L., and Berke, O. (2009). Predicting geographical human risk of West Nile Virus—Saskatchewan, 2003 and 2007. *Canadian Journal of Public Health* 100: 344–349.

Estrada-Peña, A. (1997). Epidemiological surveillance of tick populations: A model to predict the colonization success of *Ixodes ricinus* (Acari: Ixodidae). *European Journal of Epidemiology* 13: 573–580.

Estrada-Peña, A., Estrada-Sánchez, A., Estrada-Sánchez, D., and Fuente, J.D.L. (2013). Assessing the effects of variables and background selection on the capture of the tick climate niche. *International Journal of Health Geographics* 12(1): 43.

European Centre for Disease Prevention and Control (ECDC). (2012). West Nile fever maps. Stockholm, Sweden: ECDC. http://ecdc.europa.eu/en/healthtopics/west_nile_fever/West-Nile-fever-maps/. Accessed August 1, 2013.

Evans, R., Mullally, D.I., Wilson, R.W. et al. (1987). National trends in the morbidity and mortality of asthma in the US. Prevalence, hospitalization and death from asthma over two decades: 1965–1984. *Chest* 91(Suppl. 6): S65–S74.

Fenwick, A. (2006). Waterborne infectious diseases–Could they be consigned to history? *Science* 313: 1077–1081.

Foley, J.E., Queen, E.V., Sacks, B., and Foley, P. (2005). GIS-facilitated spatial epidemiology of tick-borne diseases in coyotes (*Canis latrans*) in northern and coastal California. *Comparative Immunology, Microbiology and Infectious Diseases* 28(3): 197–212.

Foody, G.M. (2006). GIS: Health applications. *Progress in Physical Geography* 30(5): 691–695.

Friedmann, P.D., Lemon, S.C., Stein, M.D., Etheridge, R.M., and D'Aunno, T.A. (2001). Linkage to medical services in the drug abuse treatment outcome study. *Medical Care* 39: 284–295.

Geronimus, A.T. and Bound, J. (1998). Use of census-based aggregate variables to proxy for socioeconomic group: Evidence from national samples. *American Journal of Epidemiology* 148: 475–486.

Gibbs, S.E.J., Wimberly, M.C., Madden, M., Masour, J., Yabsley, M.J., and Stalknecht, D.E. (2006). Factors affecting the geographic distribution of West Nile virus in Georgia, USA: 2002–2004. *Vector-Borne and Zoonotic Disease* 6: 73–82.

Gohosh, D. and Guha, R. (2010). Use of genetic algorithm and neural network approaches for risk factor selection: A case study of West Nile virus dynamics in an urban environment. *Computers, Environment and Urban Systems* 34: 189–203.

Gohosh, D. and Guha, R. (2011). Using neural network for mining interpretable relationships of West Nile risk factors. *Social Science & Medicine* 72: 418–429.

Goodchild, M.F. (1986). *Spatial Autocorrelation. CATMOG-Concepts and Techniques in Modern Geography.* Norwich, England: Geo Books, p. 47.

Gosselin, P., Lebel, G., Rivest, S., and Douville-Fradet, M. (2005). The integrated system for public health monitoring of West Nile virus (ISHM-WNV): A real-time GIS for surveillance and decision-making. *International Journal of Health Geographics* 4: 21.

Gould, P.R. (1995). *The Coming Plaque.* New York: Blackwell.

Griffith, D.A. (2005). A comparison of six analytical disease mapping techniques as applied to West Nile virus in the conterminous United States. *International Journal of Health Geographics* 4: 18–26.

Gruenewald, P.J., Freistler, B., Remer, L., LaScala, E.A., and Treno, A.J. (2006). Ecological models of alcohol outlets and violent assaults: Crime potentials and geospatial analysis. *Addiction* 101: 666–677.

Guerra, M., Walker, E., Jones, C. et al. (2002). Predicting the risk of Lyme disease: Habitat suitability for *Ixodes scapularis* in the North Central United States. Faculty Publications: Department of Entomology [Internet]. http://digitalcommons.unl.edu/entomologyfacpub/236. Accessed February 12, 2012.

Harries, K. (1995). The ecology of homicide and assault: Baltimore City and County: 1989–91. *Study of Crime Prevention* 4(1): 44–60.

Havelaar, A.H. (1994). Application of HACCP to drinking water supply. *Food Control* 5(3): 145–152.

Hay, S.I., Omumbo, J.A., Craig, M.H., and Snow, R.W. (2000). Earth observation, geographic information systems and *Plasmodium falciparum* malaria in Sub-Saharan Africa. *Advances in Parasitology* 47: 173–215.

Hernandez-Jover, M., Roche, S., and Ward, M.P. (2013). The human and animal health impacts of introduction and spread of an exotic strain of West Nile virus in Australia. *Preventive Veterinary Medicine* 109: 186–204.

Homa, D.M., Mannino, D.M., and Lara, M. (2000). Asthma mortality in U.S. Hispanics of Mexican, Puerto Rican, and Cuban heritage, 1990–1995. *American Journal of Respiratory and Critical Care Medicine* 161: 504–509.

HUD. (2015). US Department of Housing and Urban Development (HUD). http://www.census.gov/housing/. Accessed December 7, 2015.

Hughes, S., Syed, Q., Woodhouse, S., Lake, I., Osborn K., Chalmers, R.M., and Hunter, P.R. (2004). Using geographic information system to investigate the relationship between reported cryptosporidiosis and water supply. *International Journal of Health Geographics* 3: 15.

IPCC. (2000). Summary for policymakers. A special report on Emission Scenarios. Prepared by Working Group III of the Intergovernmental Panel on Climate Change, N. Nakicenovic, O. Davidson, G. Davis et al. (Eds.). Cambridge, U.K.: Cambridge University Press, 20pp.

Jacob, B.G., Burkett-Cadena, N.D., Luvall, J.C., Parcak, S.H., McClure, J.W., Estep, L.K., Hill, G.E., Cupp, E.W., and Novak, R.J. (2010). Developing GIS-based Eastern Equine Encephalitis vector-host models in Tuskegee, Alabama. *International Journal of Health Geographics* 9: 12.

Jacob, B.G., Chadee, D.D., and Novak, R.J. (2011). Adjusting second moment bias in eigenspace using Bayesian empirical estimators, Dirichlet tessellations and Worldview I data for predicting *Culex quinquefasciatus* habitats in Trinidad. *Journal of Geographic Information Systems* 3: 18–49.

Jacob, B.G., Gu, W., Muturi, E.J., Caamano, E.X., Morris, J.M., Lampman, R., and Novak, R.J. (2009). Developing operational algorithms using linear and non-linear least squares estimation in Python® for identification of *Culex pipiens* and *Culex restuans* aquatic habitats in a mosquito abatement district (Cook County, Illinois). *Geospatial Health* 3: 23–31.

Jacobson, J. (2004). Place and attrition from substance abuse treatment. *Journal of Drug Issues* 34: 23–49.

Jenness, J. (2003). Mahalanobis distances (mahalanobis.avx) extension for ArcView 3.x, Jenness Enterprises. http://www.jennessent.com/arcview/mahalanobis.htm. Accessed May 6, 2008.

Joe, G.W., Simpson, D.D., and Sells, S.B. (1994). Treatment process and relapse to opioid use during methadone maintenance. *American Journal of Drug and Alcohol Abuse* 20: 173–197.

Jones, C.G., Ostfeld, R.S., Richard, M.P., Schauber, E.M., and Wolff, J.O. (1998). Chain reactions linking acorns to gypsy moth outbreaks and Lyme disease risk. *Science* 279(5353): 1023–1026.

Jones, W., McDonald, P., Enrigth, D., and Sampson, L. (2003). Using spatial statistics to describe HIV morbidity in 3 North Carolina Counties. *National HIV Prevention Conference*, Atlanta, GA, Abstract No. TP-087.

Kelly, R.R., Gaines, D., Gilliam, W.F., and Brinkerhoff, R.J. (2014). Population genetic structure of the Lyme disease vector *Ixodes scapularis* at an apparent spatial expansion front. *Infection, Genetics and Evolution* [Internet]. http://www.sciencedirect.com/science/article/pii/S1567134814001919. Accessed August 21, 2014.

Kim, M., Holt, J.B., Eisen, R., Padgett, K., Reisen, W.K., and Croft, J. (2011). Detection of swimming pools by geographic object-based image analysis to support West Nile virus control efforts. *Photogrammetric Engineering and Remote Sensing Journal* 77: 1169–1179.

Kitron, U. (2000). Risk maps: Transmission and burden of vector-borne diseases. *Parasitology Today* 16: 324–325.

Krause, N. (1996). Neighborhood deterioration and self-rated health in later life. *Psychology and Aging* 11: 1342–1352.

Krieger, N., Chen, J.T., Waterman, P.D., Rehkopf, D.H., and Subramanian, S.V. (2003). Race/ethnicity, gender and monitoring socioeconomic gradients in health: A comparison of area-based socioeconomic measures—The public health disparities geocoding project. *American Journal of Public Health* 93(10): 1655–1671.

Krieger, N., Chen, J.T., Waterman, P.D., Soobader, M., Subramanian, S.V., and Carson, R. (2002). Geocoding and monitoring of US socioeconomic inequalities in mortality and cancer incidence: Does the choice of area-based measure and geographic level matter? *American Journal of Epidemiology* 156: 471–482.

Krieger, N.G.D. (1999). Re: Use of census-based aggregate variables to proxy for socioeconomic group: Evidence from national samples. *American Journal of Epidemiology* 150: 892–896.

Latkin, C., Glass, G.E., and Duncan, T. (1998). Using geographic information systems to assess spatial patterns of drug use, selection bias and attrition among a sample of injection drug users. *Drug and Alcohol Dependence* 50: 167–175.

Law, D.C.G., Serre, M.L., Christakos, G., Leone, P.A., and Miller, W.C. (2004). Spatial analysis and mapping of sexually transmitted diseases to optimize intervention and prevention strategies. *Sexually Transmitted Infections* 80: 294–299.

Lawson, A.B. and Kleinman, K. (2005). *Spatial and Syndromic Surveillance for Public Health*. Chichester, England: John Wiley & Sons.

Leblond, A., Sandoz, A., Lefebvre, G., Zeller, H., and Bicout, D.J. (2007). Remote sensing based identification of environmental risk factors associated with West Nile disease in horses in Camargue, France. *Preventive Veterinary Medicine* 79: 20–31.

Lewin, S., Norman, R., Nannan, N., Thomas, E., and Bradshaw, D. (2007). Estimating the burden of disease attributable to unsafe water and lack of sanitation and hygiene in South Africa in 2000. *South African Medical Journal* 97: 755–762.

Li, J., Kolivras, K.N., Hong, Y., Duan, Y., Seukep, S.E., Prisley, S.P., Campbell, J.B., and Gaines, D.N. (2014). Spatial and temporal emergence pattern of Lyme disease in Virginia. *American Journal of Tropical Medicine and Hygiene* 91: 1166–1172.

Lian, M., Warner, R.D., Alexander, J.L., and Dixon, K.R. (2007). Using geographic information systems and spatial and space-time scan statistics for a population-based risk analysis of the 2002 equine West Nile epidemic in six contiguous regions of Texas. *International Journal of Health Geographics* 6: 42. doi:10.1186/1476-072X-6-42.

Linard, C., Lamarque, P., Heyman, P., Ducoffre, G., Luyasu, V., Tersago, K., Vanwambeke, S.O., and Lambin, E.F. (2007). Determinants of the geographic distribution of Puumala virus and Lyme borreliosis infections in Belgium. *International Journal of Health Geographics* 6: 15. doi:10.1186/1476-072X-6-15.

Liu, A., Lee, V., Galusha, D., Slade, M.D., Diuk-Wasser, M., Andreadis, T., Scotch, M., and Rabinowitz, P.M. (2009). Risk factors for human infection with West Nile virus in Connecticut: A multi-year analysis. *International Journal of Health Geographics* 8: 67. doi:10.1186/1476-072X-8-67.

Liu, H., Weng, Q., and Gaines, D. (2008). Spatio-temporal analysis of the relationship between WNV dissemination and environmental variables in Indianapolis, USA. *International Journal of Health Geographics* 7: 66. doi:10.1186/1476-072X-7-66.

Liu, H., Weng, Q., and Gaines, D. (2011). Geographic incidence of human West Nile virus in northern Virginia, USA, in relation to incidence in birds and variations in urban environment. *Science of the Total Environment* 409: 4235–4241.

Maantay, J. (2007). Asthma and air pollution in the Bronx: Methodological and data considerations in using GIS for environmental justice and health research. *Health and Place* 13: 32–56.

Mahalanobis, P.C. (1936). On the generalized distance in statistics. *Proceedings of the National Institute of Sciences of India* 2: 49–55.

Mannino, D.M., Homa, D.M., Akinbami, L.J., Moorman, J.E., Gwynn, C., and Redd, S.C. (2002). Surveillance for asthma—United States, 1980–1999. *Morbidity and Mortality Weekly Report* 51(SS-1): 1–13.

Mason, M., Cheung, I., and Walker, L. (2004). Substance networks, and the geography of urban adolescents. *Substance Use and Misuse* 39(10–12): 1–27.

Mather, T.N., Nicholson, M.C., Donnelly, E.F., and Matyas, B.T. (1996). Entomologic index for human risk of Lyme disease. *American Journal of Epidemiology* 144: 1066–1069.

Matteucci, R., Schub, T., and Pravikoff, D. (2012). Sexually transmitted diseases in women. *CINAHL Nursing Guide*. Ipswich, MA: Cinahl Information Systems, pp. 1–2.

McKenna, D.F., Fausini, Y., Nowakowski, J., and Wormser, G.P. (2004). Factors influencing the utilization of Lyme disease-prevention behaviors in a high-risk population. *Journal of the American Academy of Nurse Practitioners* 16: 24–30.

McMaster, R.B., Leitner, H., and Sheppard, E. (1997). GIS-based environmental equity and risk assessment: Methodological problems and prospects. *Cartography and Geographic Information Systems* 24(3): 172–189.

McMichael, A.J. (2004). Environmental and social influences on emerging infectious diseases: Past, present and future. *Philosophical Transactions of Royal Society of London B: Biological Sciences* 359: 1049–1058.

Messina, J.P., Brown, W., Amore, G., Kitron, U.D., and Ruiz, M.O. (2011). West Nile virus in the Greater Chicago area: A geographic examination of human illness and risk from 2002 to 2006. *URISA Journal* 23: 5–22.

Mojala, S.A. and Everett, B. (2012). STD and HIV risk factors among U.S. young adults: Variations by gender, race, ethnicity, and sexual orientation. *Perspectives on Sexual and Reproductive Health* 44(2): 125–133. doi:10.1363/4412512.

Moonan, P.K., Bayona, M., Quitugua, T.N., Oppong, J., Dunbar, D., Jost, K.C., Burgess, G., Karan, P.S., and Weis, S.E. (2004). Using GIS technology to identify areas of tuberculosis transmission and incidence. *International Journal of Health Geographics* 3: 23. doi:10.1186/1476-072X-3-23.

Moran, P.A.P. (1950). Notes on continuous stochastic phenomena. *Biometrika* 37: 17–23.

Murray, C.J. and Lopez, A.D. (1997). Mortality by cause for eight regions of the world: Global Burden of Disease Study. *Lancet* 349: 1269–1276.

Nagelkerke, N.J.D., Jha, P., de V las, S.J. et al. (2002). Modelling HIV/AIDS epidemics in Botswana and India: Impact of interventions to prevent transmission. *Bulletin of World Health Organisation* 80(2): 89–96.

Nakaya, T., Nakase, K., and Osaka, K. (2005). Spatio-temporal modeling of the HIV epidemic in Japan based on the national HIV/AIDS surveillance. *Journal of Geographic Systems* 7: 313–336.

NEI. (2011). National emissions inventory. http://www3.epa.gov/ttnchie1/net/2011inventory.html.

Nicholas, H., Ogden, L., Lindsay, R., Morshed, M., Sockett, P.N., and Artsob, H. (2009). The emergence of Lyme disease in Canada. *Canadian Medical Association Journal* 180(12): 1221–1224.

Nicholson, M.C. and Mather, T.N. (1996). Methods for evaluating Lyme disease risks using geographic information systems and geospatial analysis. *Journal of Medical Entomology* 33: 711–720.

Nielsen, C.F., Armijos, M.V., Wheeler, S., Carpenter, T.E., Boyce, W.M., Kelley, K., Brown, D., Scott, T.W., and Reisen, W.L. (2008). Risk factors associated with human infection during the 2006 West Nile virus outbreak in Davis, a residential community in northern California. *American Journal of Tropical Medicine and Hygiene* 78: 53–62. http://www.ncbi.nlm.nih.gov/pmc/articles/ PMC2215055/. Accessed June 16, 2013.

Norris, D.E., Klompen, J.S.H., Keirans, J.E., and Black, W.C. (1996). Population Genetics of *Ixodes scapularis* (Acari: Ixodidae) based on mitochondrial 16S and 12S genes. *Journal of Medical Entomology* 33(1): 78–89.

Oakes, J.M. (2004). The misestimation of neighborhood effects: Casual inference for a practicable social epidemiology. *Social Science and Medicine* 58(10): 1929–1952.

Odoi, A., Martin, S.W., Michel, P., Middleton, D., Holt, J., and Wilson, J. (2004). Investigation of clusters of giardiasis using GIS and a spatial scan statistic. *International Journal of Health Geographics* 3: 11.

Ogden, N.H., Maarouf, A., Barker, I.K., Bigras-Poulin, M., Lindsay, L.R., Morshed, M.G., O'Callaghan, C.J., Ramay, F., Waltner-Toews, D., and Charron, D.F. (2006). Climate change and the potential for range expansion of the Lyme disease vector *Ixodes scapularis* in Canada. *International Journal for Parasitology* 36: 63–70.

Ogden, N.H., St-Onge, L., Barker, I.K. et al. (2008). Risk maps for range expansion of the Lyme disease vector, *Ixodes scapularis*, in Canada now and with climate change. *International Journal of Health Geographics* 7: 24.

Ostfeld, R.S., Canham, C.D., Oggenfuss, K., Winchcombe, R.J., and Keesing, F. (2006). Climate, deer, rodents, and acorns as determinants of variation in Lyme-disease risk. *PLoS Biology* 4(6): e145.

Ozdenerol, E., Bialkowska-Jelinska, E., and Taff, G.N. (2008). Locating suitable habitats for West Nile virus-infected mosquitoes through association of environmental characteristics with infected mosquito locations: A case study in Shelby County, Tennessee. *International Journal of Health Geographics* 7: 12.

Ozdenerol, E., Taff, G.N., and Akkus, C. (2013). Exploring the spatio-temporal dynamics of reservoir hosts, vectors, and human hosts of West Nile virus: A review of the recent literature. *International Journal of Environmental Research and Public Health* 10(11): 5399–5432.

Ozdenerol, E. (2015). GIS and Remote Sensing Use in the Exploration of Lyme Disease Epidemiology. *International Journal of Environmental Research and Public Health* 12(12): 15182–15203.

Pellow, D.N. (2000). Environmental inequality formation: Toward a theory of environmental justice. *American Behavioral Scientist* 43: 581–601.

Pepin, K.M., Eisen, R.J., Mead, P.S. et al. (2012). Geographic variation in the relationship between human Lyme disease incidence and density of infected host seeking *Ixodes scapularis* nymphs in the Eastern United States. *American Journal of Tropical Medicine* 86(6): 1062–1071.

Rappole, J.H., Derrickson, S.R., and Hubalek, Z. (2000). Migratory birds and spread of West Nile virus in the western hemisphere. *Emerging Infectious Diseases* 6: 319–328.

Richer, L.M., Brisson, D., Melo, R., Ostfeld, R.S., Zeidner, N., and Gomes-Solecki, M. (2014). Reservoir targeted vaccine against *Borrelia burgdorferi*: A new strategy to prevent Lyme disease transmission. *Journal of Infectious Diseases*. doi: 10.1093/infdis/jiu005.

Rodier, G. (2007). New rules on international public health security. *Bulletin of World Health Organisation* 85: 428–430.

Rogers, D.J. and Randolph, S.E. (2003). Studying the global distribution of infectious diseases using GIS and RS. *Nature Reviews Microbiology* 1: 231–237.

Rose, D., Mannino, D.M., and Leaderer, B.P. (2007). Asthma prevalence among US adults, 1998–2000: Role of Puerto Rican ethnicity and behavioral and geographic factors. *American Journal of Public Health* 96: 880–888.

Ruiz, M.O., Tedesco, C., McTighe, T.J., Austin, C., and Kitron, U. (2004). Environmental and social determinants of human risk during a West Nile virus outbreak in the greater Chicago area. *International Journal of Health Geographics* 3: 8.

SaTScan. (2013). SaTScan User Guide. http://www.satscan.org/. Accessed September 23, 2013.

Schulze, T.L., Jordan, R.A., Schulze, C.J., and Hung, R.W. (2009). Precipitation and temperature as predictors of the local abundance of *Ixodes scapularis* (Acari: Ixodidae) nymphs. *Journal of Medical Entomology* 46(5): 1025–1029.

Schwartz, B.S.L. and Goldstein, M.D. (1990). Lyme disease in outdoor workers: Risk factors, preventive measures, and tick removal methods. *American Journal of Epidemiology* 131: 877–885.

Scribner, R.A., MacKinnon, D.P., and Dwyer, J.H. (1995). Geographic risk of violence associated with alcohol outlets in Los Angeles County cities. Paper presented at the *CDC Symposium on Statistical Methods: Small Area Statistics in Public Health*, Atlanta, GA.

Semenza, J.C. and Menne, B. (2009). Climate change and infectious diseases in Europe. *The Lancet Infectious Diseases* 9: 365–375.

Shiode, N., Shiode S., Rod-Thatcher, E., Rana, S., and Vinten-Johansen, P. (2015). The mortality rates and the space-time patterns of John Snow's cholera epidemic map. *International Journal of Health Geographics* 14: 21. doi 10.1186/s12942-015-0011-y.

Smallman-Raynor, M.R., Cliff, A.D., and Haggett, P. (1992). *Atlas of AIDS*. Oxford, England: Blackwell.

Smart, R.G., Adlaf, E.M., and Walsh, G.W. (1994). Neighborhood socioeconomic factors in relation to student drug use and programs. *Journal of Child and Adolescent Substance Abuse* 3(1): 37–46.

Smith, K.F., Sax, D.F., Gaines, S.D., Guernier, V., and Guegan, J.F. (2007). Globalization of human infectious disease. *Ecology* 88: 1903–1910.

Snapinn, K.W., Holmes, E.C., Young, D.S., Bernard, K.A., Kramer, L.D., and Ebel, G.D. (2007). Declining growth rate of West Nile virus in North America. *Virology* 81: 2531–2534.

Soverow, J.E., Wellenius, G.A., Fisman, D.N., and Mittleman, M.A. (2009). Infectious disease in a warming world: How weather influenced West Nile virus in the United States (2001–2005). *Environmental Health Perspective* 117: 1049–1055.

Strachan, D.P. (2000). The role of environmental factors in asthma. *British Medical Bulletin* 56(4): 865–882.

Sugumaran, R., Larson, S.R., and DeGroote, J.P. (2009). Spatio-temporal cluster analysis of county-based human West Nile virus incidence in the continental United States. *International Journal of Health Geographics* 8: 43. doi:10.1186/1476-072X-8-43.

Tachiiri, K., Klinkenberg, B., Mak, S., and Kazmi J. (2006). Predicting outbreaks: A spatial risk assessment of West Nile virus in British Columbia. *International Journal of Health Geographics* 5(1): 1. doi: 10.1186/1476-072X-5-21.

Thomas, R. (1992). *Geomedical Systems: Intervention and Control*. New York: Routledge.

Thompson, D.R., Juarez, M., Barker, C.M., Holeman, J., Lundeen, S., Mulligan, S., Painter, T.H., Pdest, E., Seidel, F.C., and Ustinov, E. (2013). Airborne imaging spectroscopy to monitor urban mosquito microhabitats. *Remote Sensing Environment* 137: 226–233.

Tobler, W. (1979). Cellular geography. In S. Gale and C. Olsson (Eds.), *Philosophy in Geography*. Lanham, MD: Rowman & Littlefield Publishers, Inc.

Tucker, J.A., Vuchininch, R.E., and Gladsjo, J.A. (1991). Environmental influences on relapse in substance use disorders. *International Journal of Addiction 1990–1991* 25: 1017–1050.

UNAIDS. (2010). UNAIDS/WHO: AIDS epidemic update 2010: UNAIDS fact sheet. http://www.unaids.org/documents/20101123_FS_SSA_em_en.pdf. Accessed July 23, 2012.

Van Oers, J.A. and Garretsen, H. (1993). The geographic relationship between alcohol use, bars, liquor shops and traffic injuries in Rotterdam. *Journal of Studies on Alcohol* 54(6): 739–744.

Vollmer, S.A., Feil, E.J., Chu, C.-Y., Raper, S.L., Cao, W.-C., Kurtenbach, K., and Margos, G. (2013). Spatial spread and demographic expansion of Lyme borreliosis spirochaetes in Eurasia. *Infection, Genetics and Evolution* 14: 147–155.

Wallace, D. and Wallace, R. (1998). *A Plaque on Your Houses: how New York was burned down and public health crumbled*. London, U.K.: Verso.

Ward, M.P. (2005). Epidemic West Nile virus encephalomyelitis: A temperature-dependent, spatial model of disease dynamics. *Preventive Veterinary Medicine* 71: 253–264.

Ward, M.P., Ramsay, B.H., and Gallo, K. (2005). Rural cases of equine West Nile virus encephalomyelitis and the normalized difference vegetation index. *Vector Borne Zoonotic Diseases* 5: 181–188.

Ward, M.P. and Scheurmann, J.A. (2008). The relationship between equine and human West Nile virus disease occurrence. *Veterinary Microbiology* 129: 378–383.

Wey, C.L., Griesse, J., Kightlinger, L., and Wimberly, M.C. (2009). Geographic variability in geocoding success for WNV cases in South Dakota. *Health and Place* 15: 1108–1114.

WHO. (2012a). Control of sexually transmitted and reproductive tract infections, and HIV/AIDS. http://www.afro.who.int. Accessed December 7, 2015.

WHO. (2012b). Initiative for vaccine research (IVR): Sexually transmitted diseases, HIV/AIDS. http://www.who.int. Accessed December 7, 2015.

WHO. (2012c). Cause-specific mortality and morbidity. http://www.who.int. Accessed December 7, 2015.

WHO. (2014). Lyme borreliosis (Lyme disease) [Internet]. WHO. http://www.who.int/ith/diseases/lyme/en/. Accessed August 25, 2014.

Wilking, H.L. and Stark, K. (2014). Trends in surveillance data of human Lyme borreliosis from six federal states in eastern Germany, 2009–2012. *Ticks and Tick-Borne Diseases* 5: 219–224.

Williams, B., Pennock-Roma, M., Suen, H.K., Magsumbol, M.S., and Ozdenerol, E. (2006). Assessing the impact of the local environment on birth outcomes: A case for HLM. *Journal of Exposure Science and Environmental Epidemiology* 17: 445–457.

Wimberly, M.C., Baer, A.D., and Yabsley, M.J. (2008b). Enhanced spatial models for predicting the geographic distributions of tick-borne pathogens. *International Journal of Health Geographics* 7(1): 15.

Wimberly, M.C., Yabsley, M.J., Baer, A.D., Dugan, V.G., and Davidson, W.R. (2008a). Spatial heterogeneity of climate and land-cover constraints on distributions of tick-borne pathogens. *Global Ecology and Biogeography* 17(2): 189–202.

Winters, A.M., Eisen, R.J., Delorey, M.J., Fischer, M., Nasci, R.S., Zielinski-Gutierrez, E., Moore, C.G., Pape, W., and John, E.L. (2010). Spatial risk assessments based on vector-borne disease epidemiologic data: Importance of scale for West Nile virus disease in Colorado. *The American Journal of Tropical Medicine and Hygiene* 82: 945–953.

Yang, K., LeJeune, J., Alsdorf, D. et al. (2012). Global distribution of outbreaks of water-associated infectious diseases. *PLoS Neglected Tropical Diseases* 6(2): e1483. doi: 10.1371/journal.pntd.0001483.

Yiannakoulias, N.W., Schopflocher, D.P., and Svenson, L.W. (2006). Modeling geographic variations in West Nile virus. *Canadian Journal of Public Health* 97: 374–379.

Young, S.G. and Jensen, R.R. (2012). Statistical and visual analysis of human West Nile virus infection in the United States, 1999–2008. *Applied Geography* 34: 425–431.

Zeman, P. and Benes, C. (2014). Peri-urbanisation, counter-urbanisation, and an extension of residential exposure to ticks: A clue to the trends in Lyme borreliosis incidence in the Czech Republic? *Ticks and Tick-Borne Diseases* 5: 907–916.

Zenilman, J., Ellish, N., Fresia, A., and Glass, G. (1999). The geography of sexual partnerships in Baltimore: Applications of core theory dynamics using a geographic information system. *Sexually Transmitted Diseases* 26(2): 75–81.

Zhu, L., Gorman, D.M., and Horel, S. (2006). Hierarchical Bayesian spatial models for alcohol availability, drug "hot spots" and violent crime. *International Journal of Health Geographics* 5: 54.

Zuber, P.L., McKenna, M.T., Binkin, N.J., Onorato, I.M., and Castro, K.G. (1997). Long-term risk of tuberculosis among foreign-born persons in the United States. *JAMA* 278(4): 304–307.

5

Children's Lead Poisoning

Children's lead poisoning remains of concern to health-care providers and public health officials worldwide. Health officials are still struck by the geographic and demographic disparities uncovered in children's lead poisoning and make a clear case that lead poisoning is a much greater problem among children in low-income countries and marginalized populations or in children living in lead-polluted sites. Even though the number of children with elevated blood lead levels (EBLLs) in the United States is decreasing, eliminating EBLLs by the year 2020 remains a goal of the U.S. Department of Health and Human Services (Healthy People 2020 Objectives 2014). The capacity to achieve this goal is conditional on the ability to develop strategies based on geographic areas (Yasnoff and Sondik 1999). In the United States, living in the poorest neighborhoods and in a neighborhood with a preponderance of pre-1950 housing accounts for "at-risk" children. Sometimes those areas overlap, but even accounting for that overlap, each factor independently and significantly heighten the risk children face. Risk factors for childhood lead poisoning such as age of housing, urban/rural status, race/ethnicity, socioeconomic status (SES), population density, renter/owner occupancy, housing value, and nutritional status have been thoroughly parsed out in childhood lead poisoning research.

Children's Exposure to Lead

Lead is a heavy metal with a low melting point. It can be easily molded and shaped and can be combined with other metals to form alloys (see key words for forms of lead). For these reasons, lead has been widespread in products as diverse as pipes, storage batteries, pigments and paints, glazes, vinyl products, weights, shot and ammunition, cable covers, and radiation shielding. Organic lead (tetra-ethyl), used in leaded gasoline, results in the release of lead into the atmosphere and soil. All forms of lead are toxic. Once lead is introduced into the environment, it persists.

For more than 35 years, health-based guidance values for lead in water, air, and the workplace, as well as human health risks arising from foodborne lead have been developed by various task groups convened by the World Health Organization and the International Program on Chemical Safety (WHO 2015).

A large body of knowledge on the effects of lead on neurobehavioral development of children at low levels of exposure has accumulated over the years. The damage lead can do to a child's developing nervous system is irreversible. In hard-hit areas, a generation of children has been at high risk for suffering symptoms such as behavioral disorders and reduced attention span and IQ scores. Children's bodies absorb lead easily, especially in the brain and central nervous system, making them highly susceptible to the effects of lead poisoning. Sources of environmental lead contamination can be difficult to pinpoint because the pathways to lead absorption are various: (1) deteriorating lead-based paint from walls, windows, and doors, (2) transportation of lead contamination to the house by other means, (3) playing with toys that contain lead, (4) absorption of leaded dust through hand-to-mouth behavior, and (5) being in polluted environment (WHO 2014). The most common pathway could be hand-to-mouth behavior especially among young children; however, it is hard to know when and how they interact with lead contamination (McDonald and Potter 1996). Exposure during childhood is thought to be brief, usually until the age of 6 (Brown and Margolis 2012); however, the side effects persist throughout life (Graff et al. 2006).

Possible sources for lead include leaded paint, lead-contaminated soil, lead in plumbing, automobile exhaust, by-products of both mining and metal working, and various consumer products (Reed 1972; Chisolm et al. 1974; Edwards 2008; Oyana and Margai 2010; Cooper 2014).

After the ill effects of lead on people's health were recognized, lead was first banned in Europe in the early 1900s (Bochynska 2014). Lead use in the United States was successively banned in paint in 1978 (U.S. Consumer Product Safety Commission 2014), in pipes in 1986 (Public Health and Welfare 2014) and in gasoline in 1995 (United States Environmental Protection Agency 2014). Environmental lead from these sources has not been completely eliminated. Houses with old pipes and paint, which contaminate the drinking water and surrounding soil, are still a significant source of lead exposure (Edwards 2008, 2014; Brown and Margolis 2012).

Although many countries have initiated programs to lower the level of lead in the environment, and the trend over time has been that lead poisoning cases are declining, that does not mean the problem's impact has dissipated. Despite being a preventable environmental problem, lead poisoning remains a major health threat and a persistent source of illness in the United States. Its estimated cost is $50.9 billion (Trasande and Liu 2011). Changes in federal laws to limit the use of lead reversed the increasing trend in BLLs of children in the United States between 1900 and 1975, but children aged <6 years continued to be exposed to lead (Brown and Margolis 2012). With increased understanding of lead poisoning over time, the threshold of EBLL for childhood lead poisoning has lowered from 60 to 30 µg/dL in 1975 (Preventing Lead Poisoning in Young Children: A Statement by the Centers for Disease Control, March 1975, 2014), 25 µg/dL in 1985, 10 µg/dL in 1991 (Preventing Lead Poisoning in Young Children: A Statement by the Centers for Disease Control and Prevention, October 1991, 2014), and finally 5 µg/dL in 2012 (CDC 2014). To date, no safe blood lead thresholds for the adverse effects of lead on children have been identified (Brown and Margolis 2012).

In lead poisoning studies, GIS were used in various stages from data preparation, to multivariate mapping of BLLs with their risk factors, to spatial and statistical analysis. At the data preparation stage, address geocoding is the most used tool to transfer tabular data sets, such as screened children addresses, into GIS (Guthe et al. 1992; Sargent et al. 1997; Reissman et al. 2001; Miranda et al. 2002, 2007; Krieger et al. 2003; Roberts et al. 2003; Haley and Talbot 2004; Griffith et al. 2007; Oyana and Margai 2007, 2010; Kim et al. 2008; Vaidyanathan et al. 2009; Kaplowitz et al. 2010). Various GIS functions were used for multivariate mapping of BLLs and risk factors in a limited way such as linking SES data with screened data records (Litaker et al. 2000; Joseph et al. 2005), map overlays (Laidlaw et al. 2005; Lo et al. 2012), distance calculations (Graber et al. 2010), and hyperlinks to demolition sites' photos and city maps for mapping dust-fall lead loadings (Farfel et al. 2003). More sophisticated spatial methods have also been used such as spatial clustering (Griffith et al. 1998; Oyana and Margai 2007, 2010; Mielke et al. 2011a, 2013), spatial autocorrelation (Sargent et al. 1997; Griffith et al. 1998; Haley and Talbot 2004; Oyana and Margai 2007, 2010), spatial regression (Griffith et al. 2007), and risk modeling (Sargent et al. 1997; Miranda et al. 2002; Krieger et al. 2003; Haley and Talbot 2004; Griffith et al. 2007; Oyana and Margai 2007, 2010; Kim et al. 2008; Kaplowitz et al. 2010). Implementation of GIS to better target at-risk populations, addressing geocoding methods, selection of parameters used in geocoding, and the uncertainties they presented were not included in a similar level of detail in the lead poisoning research. New GIS-based studies are needed in surveillance data management, risk analysis, lead exposure visualization, and community intervention strategies where geographically targeted and specific intervention measures are taken.

This chapter focuses on current gaps in the screening strategies and also on methods that have been most effective in examining spatial epidemiology of childhood lead exposure. This chapter also discusses additional methods in GIS-utilized lead poisoning research that provide public health practitioners and policy makers the tools to better target lead poisoning preventive interventions.

Data Collection Processes

Countries vary greatly in the degree of their blood lead data collection methods, their recognition of the children's lead poisoning as a problem, and effectiveness of their lead poisoning prevention programs. Some countries have imposed bans on certain uses of lead, and many have set environmental standards. These countries deployed robust screening and prevention programs for monitoring levels of lead in blood and the environment.

In the United Sates, data collection methods vary among states. States maintain their own child-specific databases, which contain follow-up data on children with EBLLs including data on medical treatment, environmental investigations, and potential sources of lead exposure.

At-risk children are screened by blood tests for lead exposure. A screen is defined as a blood lead test. Laboratories report results of all blood lead tests, not just elevated values, to state health departments. States determine the reporting level for blood lead tests and decide which data elements should accompany the blood lead test result. Then, they transfer the results to Centers for Disease Control and Prevention (CDC)'s national surveillance system. The CDC's Childhood Lead Poisoning Prevention Program has been compiling state surveillance data since January 1, 1997. CDC does not provide data for analysis from the national database, as data collection methods vary among states.

In May 2012, the CDC implemented the reference level of 5 μg/dL (previous reference level—10 μg/dL) to identify children who have been exposed to lead and who require case management. Even though to date no safe blood lead thresholds for the adverse effects of lead on children have been identified (Brown and Margolis 2012; National Center for Environmental Health 2014), data related to children with very low BLLs have consistently been overlooked. Address information of children with BLLs ranging from 0 to 3 μg/dL may not be reported since screening efforts have primarily focused on children with BLLs above 5 μg/dL (Betsy Shockley, Childhood Lead Poisoning Prevention Program in Shelby County, Tennessee, March 15, 2013, personal communication). These nonrandom missing data can cause misinterpretation of the geographic distribution of lead poisoning.

Since CDC started to fund states directly in 2000, state health departments have implemented surveillance systems to track and maintain statewide childhood lead poisoning data. For example, LeadTrack surveillance software adapted by Tennessee State Health department provides an automated transfer of electronic patient data and coordinates (latitude and longitude) of the residential addresses of screened children based on a commercial Geocoding Web Service (i.e., Yahoo). Some county governments continue to use local surveillance systems to manage their remediation efforts such as abated housing data while tracking elevated lead levels. Once surveillance software designates coordinates for the residents of each child to facilitate monitoring, GIS is used for mapping regional prevalence estimates and environmental and socioeconomic risk factors for lead poisoning to support public health intervention strategies. The spatial patterns observed aid health departments in their community outreach efforts for prevention. The following section offers the uncertainties of screening data collection, address geocoding methods, and data aggregation. Major difficulties, aside from the uncertainties of data aggregation parameters, are the choice of spatial analytical methods and the failure to exploit the full capabilities of GIS for data collection and more accurate modeling of spatial processes in the lead exposure analysis.

Geocoding Screened Cases

Geocoding is the procedure for data aggregation and analysis in childhood lead poisoning. Health departments are concerned with data aggregation and the choice of geographical analysis for lead poisoning prevalence mapping. Data aggregation is done for two reasons: to link socioeconomic and environmental measures to lead data and to ensure data confidentiality. At the time of testing, the primary residence of each child tested is documented. Address data are used as geographical identifier. When address information exists, its accuracy can be problematic. Due to errors in the addresses and "P.O. Box" and "rural route" style addresses, screen reports could not be geocoded. Some cases are assigned to a location of a community center where the screening took place rather than the actual residential addresses. To improve the quality of geocoding, the addresses need to be confirmed in the data collection phase in a GIS environment. Such GIS-integrated screening could eliminate spatial bias due to disparities in reporting. This integration has potential to shift the current screening practice from health-care provider to a GIS-informed screening, which would increase the case finding rate and data completeness.

There are two main types of error that occur when geocoding: (1) inaccuracy of the geocoded locations (positional accuracy) and (2) inability to geocode all the desired locations. In the former case, an error in a screened child's residential address can introduce spatial uncertainty about where she/he lives and this error will further bias any associations between risk factors and childhood lead exposure. There is a growing body of research assessing the potential bias that may result from position error including evaluation of the positional difference between geocoding methods (Duncan et al. 2011); comparison of a single-stage geocoding method to a multistage geocoding method (Lovasi et al. 2007), urban–rural variation in the accuracy (Gregorio et al. 1999; Cayo and Talbot 2003; Oliver et al. 2005; Ward et al. 2005; Hay et al. 2009), comparison of performance of different vendors' geocoding software (Ward et al. 2005; Whitsel et al. 2006), comparison of geocoding results against imagery (Cayo and Talbot 2003) or GPS data (Wu et al. 2005), benefits of manual geocode correction (Goldberg et al. 2008), comparison of user-generated farmers' market coordinates against the geocoded coordinates from multiple geocoding services (Cui 2013), and comparison of address points, parcel, and street geocoding techniques (Zandbergen 2008). Figure 5.1 shows a comparison of a geocoded location based on the three reference data models (rooftop, tax parcel, and street) by their geocoding reliability representing better positional accuracy. While in these examples, the points are relatively close together, the differences in geocoding method can result in points falling in different census blocks, census tracts, or zip codes. Very few studies have examined these aggregation problems and spatial scale effects to monitor risk factors (Griffith et al. 2007).

FIGURE 5.1
A comparison of a geocoded location based on the three reference data models (rooftop, tax parcel, and street).

In the case where all the desired locations cannot be geocoded, nongeocoded cases could be associated with particular subsets of the population such as subgroups of high-risk children that might go unnoticed. Nonrandom missing data can result from social, economic, or political reasons (Vach and Blettner 1991) and could be systematically related to lead exposure. A geographic selection bias occurs when there is a nonrandom pattern of the nongeocoded cases. This bias can result in the detection of childhood lead poisoning clusters in particular subgroups of the population while decreasing the power to detect clusters in other subgroups. There are a number of studies that address geocoding error of this nature, not directly in lead poisoning but other epidemiologic areas, including selection bias due to nongeocoded data for American Indian reservations and areas of low population density (Wey et al. 2009), selection bias due to missing address data for rural locations disfavoring less educated males in prostate cancer study (Oliver et al. 2005), selection bias in breast cancer diagnosis favoring minorities and urban residents (Gregorio et al. 1999), and misclassification of an address into an incorrect areal unit and the effect on the cancer rates by choice of interpolation method (Goldberg and Cockburn 2012).

If geocoding success is higher in certain geographic areas or in particular segments of the population, then the results of spatial/statistical analyses may reflect the pattern of geocoding success rather than the true pattern of childhood lead poisoning. Kim et al. (2008) investigated how much the additional data from more intensive geocoding processes improved performance of childhood lead exposure risk models in identifying areas of elevated lead exposure risk. They found that the overall performance of these models tends to be driven at least partly by the underlying quality of address data in both the tax parcel and the lead surveillance data sets, as well as the total number of children previously screened. Kaplowitz et al. (2010) compared predictive validity of several lead poisoning risk assessment methods. They found that using census block groups explains much more of the variance in BLL than does dichotomizing zip codes into high and low risk. Their study confirmed that one's BLL is more closely associated with characteristics of one's immediate environment than with characteristics of a larger area, such as a tract or zip code. Kriger et al. (2003) examined temporal and spatial scale effects and the choice of geographical level of area (census block group, census tract, and zip code) to monitor social inequalities in childhood lead poisoning. They found that choice of measure and level of geography matter. Census tract- and block group-level-based socioeconomic measures consistently detected equivalent and typically stronger socioeconomic gradients than their zip code-level counterparts for the lead data. They specifically suggest that records should be geocoded to the tract or block group level and be linked to easily understood poverty-related measures. Krieger et al. (2003) noted in their conclusions that U.S. studies using area-based socioeconomic measures reported geocoding a high proportion of cases to the census tract and block group level, but none of the studies provided information on the accuracy of their geocoding.

Screening Techniques

Children's blood lead levels are collected as part of screening programs. Screening interventions are designed to identify lead poisoning in a community early, thus enabling early identification, treatment, and eradication of childhood lead poisoning. Several types of screening exist: voluntary screening based on fewer children screened, which is less effective; mandatory screening, in which more children are screened and more concentration of cases are detected; universal screening involves screening of all children in a certain category (e.g., all children of a certain age); targeted screening based on risk-assessment focus on children who may be at risk. Cost–benefit analyses do not justify universal screening in some communities with low lead exposure. Conversely, many children with elevated BLLs still are not being identified and followed up because of inadequate screening efforts in their communities. To ensure that prevention approaches are appropriate to local conditions, and to reduce unnecessary testing of children unlikely to be exposed to lead, appropriate screening policies should be implemented with GIS-assisted public health intervention strategies.

The deduplication process of screened data is also an important process that affects the spatial analysis results. The blood lead data, resulting from analysis of both capillary and venous blood specimens, are collected by a variety of public and private laboratories and forwarded to a statewide database. States maintain their own child-specific databases so they can identify duplicate test results or sequential test results on single children. This process is called the deduplication process. The deduplication reduces the data set, with each record representing the test result obtained from a single BLL screening per child or the first value obtained for those with multiple screenings or highest blood level. Some studies (Sargent et al. 1997; Mielke et al. 2011b) used first screening test for deduplication, whereas others (Miranda et al. 2002, 2007, 2011; Haley and Talbot 2004; Miranda and Dolinoy 2005; Kim et al. 2008) used the screening result with the highest BLL. Apart from these two approaches, Oyana and Margai study (Oyana and Margai 2010) used the average of test results with multiple screenings. Figure 5.2 shows a comparison of clustered areas of EBLLs in children using the first and last screening test results for a 20-year period between 1994 and 2014 in Shelby County, TN. Last screening maps may represent an improvement in intervention efforts.

In the United States, the CDC issued guidelines, endorsing universal screening in areas with ≥27% of housing built before 1950 and in populations in which the percentage of 1- and 2-year-olds with elevated BLLs is ≥12% (CDC 1997; Committee on Environmental Health 1998). For children living in areas where universal screening is not warranted, the CDC recommends

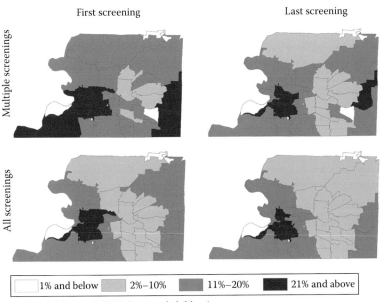

(Rate: Children with EBLL/screened children)

FIGURE 5.2
Elevated blood lead level rates among children by zip code in Shelby County, Tennessee, 1994–2014. (Courtesy of Memphis and Shelby County Health Department, Memphis, TN.)

targeted screening based on risk assessment during specified pediatric visits. In 2004, the CDC updated its screening guidelines and provided a basis for public health authorities to decide on appropriate screening policy using local BLL data (CDC 2005). This strategy was intended to increase the screening and follow-up care of children who most needed these services and to ensure that prevention approaches were appropriate to local conditions. Due to this policy in 2004, many health departments shifted their approach from universal screening to targeted screening based on census parameters. A few continued door-to-door screening. Census parameters of median income, number of family members in the household, and previous lead poisoned children locations are used to identify high-risk areas. Delineation of these high-risk areas helps local public health officials to do door-to-door testing, screening, and educational activities at local organizations, churches, and community centers. In addition to these high-risk areas, the well child prescreening questionnaire administered by clinics and the health departments also define who needs to be tested.

The next section explores more explicitly studies of screening methodology design along with GIS and spatial methods applied. This review covers a 21-year period that includes GIS-based studies published since 10 µg/dL thresholds were first introduced in 1991 until the new threshold of 5 µg/dL in 2012. Reviewed articles were summarized and grouped into five categories: screening methodology design, risk modeling studies, environmental risk factors, spatial analysis of genetic variation, and political ecology. Table 5.1 presents these studies under each category with GIS methods applied, study region, and common risk factors or major findings. The first three categories focus on children's environment. The fourth category, spatial analysis of genetic variation, focuses on individual's traits. The last category, political ecology, focuses more on the long-term socioeconomic process of childhood lead poisoning.

Spatial Analytical Methods

Screening Activities

Studies on childhood lead poisoning surveillance that used GIS include Lutz et al. (1998), Reissman et al. (2001), Roberts et al. (2003), and Vaidyanathan et al. (2009). These studies followed CDC's guidance on targeted screening (CDC 2014). The guidance requires that children at ages of 1 and 2 or ages of 3 and 6 should be tested if they have not been tested before and fall in at least one of the following criteria: residing in a ZIP code in which ≥27% of housing was built before 1950; receiving public assistance from programs such as Medicaid or the Special Supplemental Nutrition Program for Women, Infants,

TABLE 5.1

Summary of Lead Studies with Common Risk Factors and Major Findings

GIS Analysis/Citation	Region/Date	Common Risk Factors/Major Findings
Screening methodology design		
Overlay analysis, choropleth mapping (Lutz et al. 1998)	Knoxville, TN/1998	Old housing, and proximity to old roads. The screening data based on the study's risk criteria thoroughly represent the targeted population.
Address geocoding, overlay analysis, choropleth mapping (Reissman et al. 2001)	Jefferson, KY/2001	Old housing. Percent of children with EBLLs is strongly associated with old housing. The screening data based on the study's risk criteria do not fully represent the targeted population.
Address geocoding, overlay analysis (Roberts et al. 2003)	South Carolina/2003	Old housing. EBLLs are strongly associated with old housing. The screening data based on the study's risk criteria do not fully represent the targeted population.
Address geocoding, overlay analysis, choropleth mapping (Vaidyanathan et al. 2009)	Atlanta, GA/2009	Poverty, old housing. The screening is strongly correlated with WIC (Special Supplemental Nutrition Program for WIC enrolment) status but not with old housing.
Risk modeling studies		
Spatial autocorrelation (Sargent et al. 1997)	Rhode Island/1997	Old housing, poverty, vacancy, percent screened children, and percent immigrants. Older houses and vacant housing are significantly associated with excessive childhood lead exposure.
Address geocoding, overlay analysis, choropleth mapping (Miranda et al. 2002)	Durham, NC/2002	Old housing, income, and race. The percentage of African-American population, median income, and construction year of housings are significantly associated with childhood lead exposure.
Address geocoding/ (Krieger et al. 2003)	Rhode Island/2003	Poverty, education, occupation, wealth/ BLLs are strongly associated with poverty but not education level, occupation, and wealth.
Spatial autocorrelation with Simultaneous Autoregressive Model (SAR)/(Haley and Talbot 2004)	New York/2004	Old housing, race, poverty, population density, education, vacant housing, renting, and seasonality. The age of housing, education level, and percentage of African-American population variables are significant predictors of BLLs.
Point in polygon analysis, address geocoding, and spatial regression (Griffith et al. 2007)	Syracuse, NY/2007	House value, race. EBLLs are significantly associated with the percentage of African-American population and average house value.

(Continued)

TABLE 5.1 (*Continued*)

Summary of Lead Studies with Common Risk Factors and Major Findings

GIS Analysis/Citation	Region/Date	Common Risk Factors/Major Findings
Spatial autocorrelation, kriging, Local Moran's I, and LISA (Oyana and Margai 2007)	Cook, IL/2007	Old housing, income, and minority populations. The authors concluded that the dependent variable is significantly associated with housing age, income, and minority populations.
Address geocoding, risk modeling (Kim et al. 2008)	North Carolina/2008	Old housing, race, percent Hispanic, income, poverty, and seasonality. All variables are significantly associated with childhood lead exposure.
Address geocoding, sensitivity analysis (Kaplowitz et al. 2010)	Michigan/2010	Old housing, race, poverty, race, and education. BLL is associated with children's immediate environment than a larger area such as a census tract or ZIP code.
Spatial autocorrelation, kriging, Local Moran's I, and LISA (Oyana and Margai 2010)	Cook, IL/2010	Old housing, income, and minority populations. The authors concluded that the dependent variable is significantly associated with housing age, income, and minority populations.
Environmental risk factors		
Address geocoding, choropleth mapping, and overlay analysis (Guthe et al. 1992)	New Jersey/1992	Proximity to industrial sites emitting lead and hazardous waste sites contaminated with lead, and proximity to roads with high traffic volume.
3D surface modeling (Mielke et al. 1997)	New Orleans, LA/1997	Old housing, soil lead concentration. Association found between high soil lead areas and neighborhoods where children with EBLLs reside.
Choropleth mapping, overlay analysis, kriging, spatial autocorrelation (Griffith et al. 1998)	Syracuse, NY/1998	Old housing, race, population density, house value, rent. BLLs are correlated with percentage of children at risk, population density, mean housing value, and percentage of the African-American population.
Overlay analysis, choropleth mapping (Gonzalez et al. 2002)	Mexico/2002	Proximity to a point source of lead exposure. There is a significant association between children with EBLLs and their distance to a point source of lead exposure.
Address geocoding, overlay analysis (Miranda et al. 2007)	North Carolina/2007	Old housing, race, income, seasonality, water system. There is a correlation between water treatment systems and lead exposure among children.
Overlay analysis and kriging (Mielke et al. 2011)	New Orleans, LA/2011	Proximity to old and heavily used roads. Lead additives in gasoline had more impact on childhood lead exposure than the dust from leaded paint.

(*Continued*)

TABLE 5.1 (*Continued*)

Summary of Lead Studies with Common Risk Factors and Major Findings

GIS Analysis/Citation	Region/Date	Common Risk Factors/Major Findings
Overlay analysis, buffer analysis, spatial masking (Miranda et al. 2011)	North Carolina/2011	Proximity to local airports. Significant positive association found between BLLs and the distances to the airport locations. Seasonality, age of housing, median household income, and minority neighborhoods are also associated with BLLs.
Overlay analysis and Kriging (Mielke et al. 2013)	New Orleans, LA/2013	Soil lead concentrations in the old city core. A statistically significant relationship found between BLLs and soil lead level proximity to old city cores.
Spatial analysis of genetic variation		
Choropleth mapping, overlay analysis (Miranda and Dolinoy 2005)	Durham, NC/2005	Race and genetic vulnerability.
Political ecology		
Moran's I, LISA, and spatial autocorrelation (Hanchette 2008)	North Carolina/2008	Old housing, poverty, tenant farming associated with the production of tobacco, rural African-American population distribution.

and Children (WIC); and whose parents or guardians answer "yes" or "don't know" to at least one of the questions in a basic personal-risk questionnaire.

The following questions are included in the questionnaire: "Does your child live in or regularly visit a house that was built before 1950?" "Does your child live or regularly visit a house built before 1978 with recent or ongoing renovations or remodeling within the last six months?" "Does your child have siblings or playmates who has lead poisoning?" Some states had additional questions to the CDC questionnaire. Lutz et al. (1998) defined the "at-risk" population based on the questionnaire criteria. The study identified old housing and proximity to old roads as most common risk factors among those screened children. The authors produced three maps using the questionnaire data and census demographics. One of the maps shows the percentage of positive screenings for each census tract and another one displays "at-risk" and not "at-risk" screenings overlaid with the percentage of houses built before 1950. The third map plots EBLL children with the percentage of houses built before 1950. Although the study mapped the exact location of children, the state of Tennessee and some other states recently banned the disclosure of exact locations of the subjects in compliance with the Health Insurance Portability and Accountability Act (HIPAA) guidelines (Summary of the HIPAA Privacy Rule 2014; Summary of the HIPAA Security Rule 2014). Lutz et al. found that the screening data thoroughly represent the targeted population in Knoxville, TN.

Reissman et al. (2001) used GIS to assist the health department's decision-making on screening activities in Louisville, Kentucky. The study attempts to (1) assess the efficacy of Jefferson County The Childhood Lead Poisoning Prevention (CLPP) program in surveying "at-risk" children and (2) determine the capability of GIS to find neighborhoods or housing units that pose risks to children. The first part of the study focuses on the childhood lead poisoning problem at the neighborhood level, whereas the latter part examines the problem at the household level. Different from the Lutz et al. (1998) study, Reissman et al. (2001) considered the "at-risk" population as children between 6 and 35 months of age who reside in a home built before 1950 or live in a target zone where more than 27% of houses were built before 1950. The authors compared the percentage of screened children with corresponding target zones by both census tracts and ZIP codes. The study found that the percentage of children with EBLLs is strongly associated with old housing. The study also showed that significant numbers of children who live in at-risk areas were not being tested throughout the county. The second part of the study mapped the children who are younger than 7 years old with confirmed BLL ≥20 μg/dL and the houses where more than one child resides with confirmed BLL ≥20 μg/dL.

Roberts et al. (2003) conducted a study over targeted lead-screening development using GIS in Charleston County, South Carolina. The authors obtained pediatric blood tests between 1991 and 1998 from Charleston County Lead Poisoning Prevention Program. Construction year of the houses was extracted from The Charleston County Tax Assessor. The authors first geo-coded the children BLLs and then the buildings in the tax assessor records by using Matchmaker/2000 address geocoding software. After the removal of duplicate building addresses from the tax assessor, the authors merged the two geocoded data sets: children BLLs and tax assessor buildings in Charleston County. Apart from Lutz et al. (1998) and Reissman et al. (2001), the authors categorized the housing variable into three categories, pre-1950, 1950–1977, and post-1977, in order to be consistent with the CDC's recommendations. Lead poisoning prevalence ratios in these time frames were compared. The study also displayed the actual locations of the children who have EBLLs (10 μg/dL and above). The study found that the children who live in a housing unit built before 1950 are four times more likely to have EBLLs than the children who live in a housing unit built after 1950. The study also found that there is no statistically significant difference between the children who live in a housing unit built between 1950 and 1977, and those who live in a housing unit built after 1977. In terms of screening activities, the study found that some areas with a high number of pre-1950 housing were not screened at all.

Vaidyanathan et al. (2009) developed a methodology to assess neighborhood risk factors for lead poisoning problems in Atlanta, GA, in 2009. Unlike the studies referred to in the section earlier, this study primarily used the Special Supplemental Nutrition Program for WIC enrollments to identify "at-risk" populations. The authors used BLL data of children younger than

3 years of age when their blood was drawn in 2005. Three data sets were used in the study: pediatric blood tests by the Childhood Lead Poisoning Prevention Program in Georgia, the land parcel data set for 1999 by the Center for GIS at the Georgia Institute of Technology in Atlanta, and census block-group-level data from the 2000 U.S. Census data set. Since the boundary of block groups and neighborhoods did not coincide, the study followed a GIS methodology to transfer the age demographics from block groups to the neighborhood level to integrate residential land parcel data and blood lead tests with the demographics at the neighborhood level. The study indicated that only 11.9% of children aged ≤36 months from the city of Atlanta were tested for lead poisoning despite the risk of high lead exposure. The authors created a lead exposure index for the neighborhoods based on housing age and poverty. The poverty measure was calculated based on the number of children who were enrolled to the WIC. Housing age risk levels were composed of pre-1950 and pre-1978. The study reveals that 90% of residential units in Atlanta were built before 1978. These housing units might be an important source of lead exposure since most studies in the literature established a relationship between old housing and lead exposure through leaded paint. The study found that some neighborhoods are having as low as 8% of testing in children for lead poisoning whereas more than 78% of the children lived in housing units built before 1950. Excluding the Lutz et al. (1998) study, all of the studies in this section demonstrate that corresponding health departments failed to account for "at-risk" populations. The studies also demonstrate that GIS could be an effective tool to target "at-risk" neighborhoods by health departments.

Risk Modeling

This section refers to nine articles on risk model development for childhood lead poisoning (Sargent et al. 1997; Miranda et al. 2002; Krieger et al. 2003; Haley and Talbot 2004; Griffith et al. 2007; Oyana and Margai 2007, 2010; Kim et al. 2008; Kaplowitz et al. 2010). Sargent et al. (1997) conducted a census tract analysis of childhood lead exposure in Rhode Island. The study used 17,956 BLL screening records from the children who were aged 0–59 months and who were screened between 1992 and 1993. Due to the small area problem, the authors excluded two of the census tracts where there were very few screening samples. The study used the percentage of children with BLL ≥10 μg/dL as the dependent variable. The population of children for the census tracts was assigned based on census estimates. The study's final model includes five independent variables that explained 83% of the variance in lead exposure. According to the final model, households with public assistance income, houses built before 1950, vacant houses, and recent immigrants are positively associated with the outcome measure. Percentages of houses built before 1950 and vacant houses are significantly associated with the dependent variable. The source of lead exposure in immigrant children was unknown due to the possibility that they could have been exposed to lead in their home countries. The study also found that there is no association between the percentage of African-American children and high lead exposure in Rhode Island.

Miranda et al. (2002) used a tax level address geocoding procedure to show high-risk areas for the North Carolina Childhood Lead Poisoning Prevention Program. The study covered the following North Carolina counties: Buncombe, Durham, Edgecombe, New Hanover, Orange, and Wilson. The authors first geocoded the screened children at the tax parcel unit to detect the age of housing from tax assessors data sets. Overall geocoding match rates vary from 47.2% to 72.1% for the six counties in North Carolina. Using this geocoded data set, the authors employed analysis of variance (ANOVA) and multivariate analysis to determine whether the independent variables (age of the building, median income, and race) are statistically associated with the BLLs. Miranda et al. also prioritized four risk areas in Durham, NC: (1) predicted parcels that are most likely to contain leaded paint, (2) predicted parcels that are less likely to contain leaded paint, (3) predicted parcels that are lesser likely to contain leaded paint, and (4) predicted parcels that are least likely to contain leaded paint. Unlike the Sargent et al. (1997) study, Miranda et al. (2002) found that the dependent variable is correlated with the percentage of African-Americans as well as median income and construction year of housings. One major shortcoming of the model is missing data since address geocoding rates may be under 50%. This study was later updated by Kim et al. (2008). The authors investigated how much the additional data from more intensive geocoding processes improved performance of childhood lead exposure risk models in identifying areas of elevated lead exposure. They used a comprehensive three-level stepwise address geocoding process. Similar to the studies by Miranda et al. (2002), Griffith et al. (2007), and Kim et al. (2008) also deployed an address geocoding based on the cadastral parcel reference system. Also similar to the Miranda et al. (2002) study, the geocoding success rate was lower because 31.2% of the addresses were not geocoded. The results in the Kim et al. (2008) support the findings of the Miranda et al. (2002) study and also find support for the following independent variables: percentage of Hispanic population, percentage of households with public assistance, and seasonality, which are also strongly associated with BLLs in the studied population.

Kriger et al. (2003) examined temporal and spatial scale effects and the choice of geographical unit (i.e., census block group, census tract, and ZIP code) to monitor social inequalities in childhood lead poisoning. The authors used blood lead level screenings of children who live in Rhode Island. The screening period was between 1994 and 1996. Different from Miranda et al. (2002), Kriger et al. (2003) used a street reference system (known as Topologically Integrated Geographic Encoding and Referencing [TIGER] data set) for their address geocoding process. Street reference systems generally produce higher geocoding success rates compared to cadastral parcel reference systems.

For instance, the Kriger et al. (2003) study produced more than 90% of geocoding success in all geographic units, census block groups, census tracts, and ZIP codes. However, one potential weakness of the method is that street geocoding results may be distant from the actual location of houses since the method uses a linear interpolation on street segments in the reference file. The authors found that the choice of measure and the level of geography matter. Census tract and census block group socioeconomic measures detected stronger socioeconomic gradients than the zip code units. The results indicate that BLLs are strongly associated with poverty but not education level, occupation, and wealth. A similar sensitivity analysis was conducted by Kaplowitz et al. (2010). Kaplowitz et al. assessed predictive validity of different geographic units for their risk assessment. According to their study, census block groups explain more variance in BLL than high- and low-risk ZIP codes. Their study confirmed that children's BLL is more closely associated with characteristics of their immediate environment than with characteristics of a larger area such as a census tract or ZIP code.

Haley and Talbot (2004) presented a spatial analysis of BLLs in New York for the children born between 1994 and 1997. The study used the highest test result when there are multiple screens for a child. Since the BLL records contain ZIP codes for the children, the authors used ZIP codes as the geographic units for spatial analysis. Based on previous published studies, Haley and Talbot selected the following socioeconomic variables: the percentage of houses built before 1950, the percentage of adults ≥25 years of age who did not receive a high school diploma, the percentage of children living below the poverty level, the percentage of vacant housing units, the percentage of the population that rents a home, the percentage of the population screened in summer (July–September), population density, and the percentage of African-American births. The authors also used GIS to distribute the socioeconomic data proportionally to the ZIP codes and to find the centroid locations of census blocks. To deal with missing data in the lead database, the authors used the mother's race from birth certificates and estimated the proportion of African-American children for each ZIP code area. Unlike Sargent et al. (1997), this study used a different methodology to deal with the small area problem. Using GIS, the authors merged the ZIP code areas when they had less than 100 screened children. Percentage of children with EBLLs in each ZIP code was defined as the dependent variable in the statistical analysis. The authors ran a multiple linear regression analysis to identify the relationship between the BLLs and the explanatory variables. They also analyzed the residuals' spatial autocorrelation in the model using SpaceStat software and developed a simultaneous autoregressive model (SAR). Their regression analysis indicates that the age of housing, education level, and percentage of African-American population variables are significant predictors of BLLs.

Griffith et al. (2007) conducted an address geocoding study in 2007. The authors used BLLs data of children in Syracuse, NY between 1992 and 1996.

The study compares two different address geocoding methods to find the impact of positional accuracy on spatial regression analysis of children's BLLs. These geocoding methods are based on street or polygon reference systems. The Haley and Talbot data referred to earlier used ZIP code boundaries as the polygon reference system. Griffith et al. (2007), on the other hand, used cadastral parcels as the reference files. Geocoding success rate is generally much higher with street reference files than ones with cadastral parcel reference files. However, cadastral parcel reference files provide more precise geocoding results and the construction year of housing units. The authors compared cadastral and TIGER-based geocoded addresses in three sections including, census tract, census block group, and census blocks of 1990 and 2000 census demographics. The study shows that there is a noticeable but not considerably high positional error difference in their spatial statistical analyses using the two methods. The regression analysis in the study was employed in two different BLL thresholds, 5 and 10 µg/dL. Regardless of the threshold level, the results indicate that EBLLs are significantly associated with the percentage of the African-American population and average house value in the census block and census block group analyses.

Using descriptive discriminant and odds ratio analyses, Oyana and Margai (2007, 2010) created a profile of high-risk areas based on housing age, the SES, and ethnicity of the population of Chicago. The purpose of the study was to identify the health disparity among children who have different racial makeup. The study also assesses the spatiotemporal dynamics of the disease and identifies the socioeconomic and racial composition of high-risk communities in Chicago. In addition, two different types of blood test methods (capillary and venous) were compared to one another for the BLL over 10 µg/dL. Oyana et al. uses a GIS scripting tool to deduplicate pediatric blood data. This study also differs from others by producing a kriging map for the area. The kriging map of Chicago shows that Westside area has the highest risk of EBLLs in the city. The authors also used TerraSeer's Space-Time Intelligence Systems to explore the kriged prevalence rates in order to analyze spatial patterns (Jacquez 2010). Moran's I (1950) and Local Indicators of Spatial Association (LISA) statistics (Anselin 1995) were used with spatial autocorrelation to show the spatial patterns and health disparities in childhood lead toxicity in Chicago. The variations in raw prevalence rates for BLLs were high. However, kriging reduced the variations dramatically. The authors concluded that the dependent variable is significantly associated with housing age, income, and minority populations.

Environmental Risk Factors

This section discusses eight studies that address environmental risk factors (Guthe et al. 1992; Mielke et al. 1997, 2011a, 2013; Griffith et al. 1998; Gonzalez et al. 2002; Miranda et al. 2007, 2011). Guthe et al. (1992) conducted one of the earliest GIS studies on childhood lead poisoning in 1992. The authors studied New Jersey municipalities of Newark, East Orange, and Irvington. The study mapped blood screening records overlaid with census tracts in the municipalities. Children blood samples were from the years 1983 to 1990. Unlike all the studies reviewed in this article, their study used a 15 µg/dL threshold level, which was the BLL threshold level at the time. This study used street-level address geocoding. Guthe et al. used command line address matching software, which is one of the oldest address geocoding engines. In terms of environmental factors, Mielke et al. (1997) studied the associations between childhood BLLs and soil lead in Louisiana. The study used three data sets: soil lead data, age of housing data, and children blood lead data for urban New Orleans and rural Lafourche Parish in Louisiana. The study focused on soil contamination and leaded paint sources of lead toxicity problems. The percentage of housing built before 1940 was considered an indicator of leaded paint. Using x and y coordinates of census tract centroids, the authors plotted the three data sets within a three-dimensional spatial model. The study showed that there is an association between high soil lead areas and neighborhoods where children with EBLLs reside. The study suggests that inner-city children should be the focus area to eliminate lead toxicity problems in the population.

Griffith et al. (1998) employed several GIS tools that include geocoding, buffer analysis, and interpolation techniques such as kriging to depict the lead poisoning problem in Syracuse, NY. This study shows the geographic distribution of lead toxicity in Syracuse, NY in three aggregated levels: census block, census block group, and census tract. Linear regression with spatial autocorrelation is used as a statistical method for the three aggregated levels. The study shows that there is a major difference between urban and rural exposure, which is consistent with the results from Laidlaw and Filipelli (2008) and Mielke et al. (2010, 2011). It however finds no statistically significant relationship between historically heavily traveled streets and lead exposure. Lead poisoning is detectable regardless of the level of geographic resolution.

Gonzalez et al. (2002) investigated the possible impact of point sources of lead exposure relative to other types of lead exposure sources. The study was conducted in Tijuana, Mexico with Hispanic children aged between 1.5 and 6.9 years. To deal with the confounding variable of cultural habits, the study used BLLs where the subjects reported that they did not use lead-glazed ceramics for cooking or food storage purposes. The study was composed of 76 samples from 14 sites. Gonzalez et al. mapped the distribution of these 76 point sources as well as five point sources containing 19 soil samples with the values ranging from 100 to 7870 µg/g soil lead. They compared the children BLLs with Bocco and Sanchez (1997) study's prediction model, which was based on fixed industrial lead point sources. Similar to the Bocco and Sanchez study, the authors assigned Tijuana census tracts the labels of "high," "medium," "low," and "N/A" risk levels based on proximity to the lead point sources. The authors also mapped these risk levels of census tracts and children cases with EBLLs ($\geq 10 \ \mu g/dL$) where the subjects reported nonuse of lead-glazed ceramics.

Miranda et al. (2007) explored the potential effect of the use of chloramines in water treatment systems over childhood lead exposure in Wayne County, NC. The authors examined the relationship between these potential effects and the age of housing to help guide policy practices in North Carolina. The authors used the data sets of children BLLs, tax parcels, census data, and water treatment system boundaries. Children BLLs were geocoded based on tax parcels with a 72.4% geocoding success rate from the surveillance data between 1999 and 2003. The study used multivariate regression to analyze the data and concluded that the use of chloramines in the water treatment systems might inadvertently increase lead exposure among children.

Another environmental study by Miranda et al. (2011) investigated the relationship between avgas lead exposure and children BLLs. The authors selected 66 airports in 6 counties of North Carolina based on the availability of tax assessor data, the volume of air traffic, and the number of screened children for lead toxicity. The study used the airports' estimated annual lead emissions that were obtained from the U.S. EPA Office of Transportation and Air Quality. The children BLL data composed of the blood tests conducted between 1995 and 2003 for the children between the ages of 9 months and 7 years. The authors determined the airport boundaries using tax parcel data and created buffer zones surrounding each airport. The buffers were created based on the distances of 500, 1000, 1500, and 2000 m from the polygon edges of the airports. Unlike most of the studies discussed in this review article, Miranda et al. used GIS to show children locations in a jittered representation even though they ran the statistical model based on actual point locations. Using the geocoded locations, Miranda et al. were able to join children locations and buffer zones, which were created from the airport boundaries. The authors assigned dummy variables to children locations based on the boundaries mentioned earlier and seasons for the screening time. The model includes the age of housing, screening season, and demographic variables. The authors also used inverse population weights to eliminate the possible bias caused by high numbers of screening cases on parcels. The study found a significant positive association between logged BLLs and the distances to the airport locations. It further showed that seasonality is an important factor in estimating BLLs. In fall, spring, and summer seasons, children were found having higher BLLs on average compared to winter season screenings. Age of housing was negatively associated with BLLs while the median household income and minority neighborhoods had positive associations with BLLs.

Mielke et al. (2011) conducted a comparative analysis of lead poisoning problems by assessing the pre-Katrina blood and soil lead concentrations around public and private properties in New Orleans. Soil lead data were composed of 587 soil samples (224 samples from public properties and 363 samples from residential private properties) and 55,551 BLL screening records for the years between 2000 and 2005. The study showed significant differences among the blood lead prevalence between the inner city and outlying areas of New Orleans. The study also found no other statistically significant differences between inner and outer cities. The authors found that, among the screens in public properties, differences between inner and outer cities in lead toxicity prevalence are a better proxy than age of construction. The study noted that lead additives in gasoline had more impact on childhood lead exposure than the dust from leaded paint. In terms of lead dust from vehicles, the largest amount of lead was deposited on soil in the inner cities whereas outer cities were not experiencing a large amount of lead deposit from the exhaust due to a lighter traffic volume. Consequently, the study indicated that lead toxicity originated from soil contamination could help explain lead toxicity in children.

Mielke et al. (2013) analyzed the association between children blood lead levels and soil lead concentrations in relation to before and after hurricane Katrina in New Orleans. In the study, pre-Katrina was from 2000 to 2005 and post-Katrina was from 2006 to 2008. Children's blood samples (55,551 records in pre-Katrina and 7,384 records in post-Katrina period) were geocoded at the 1990 census tract level. Soil lead data were composed of 5467 post-Katrina soil samples. Soil samples were categorized by their 1 m proximity to "busy streets," "residential streets," "house sides," and "open spaces." Census tract medians of soil lead concentration data were used to produce kriging maps of soil lead concentration for both pre- and post-Katrina periods. Census tracts were also categorized as low and high in lead concentration groups based on a 100 mg/kg threshold ($\geq$100 mg/kg and <100 mg/kg). Nonparametric statistics were used because of positive skewness in the soil lead data. Multipurpose permutation procedure showed that there was a significant difference between low and high lead tracts. This confirms the significance of 100 mg/kg as a threshold for lead concentration in soil for New Orleans. Census tract soil lead concentration medians showed that busy streets had the highest median lead concentration by location. This could be related to historical lead deposits from car exhausts. Kriging maps showed that there was no major change in the lead concentration level in soil for pre- and post-Katrina periods. Unlike Griffith et al. (1998), the Mielke et al. (2013) study suggests that there is a statistically significant relationship between BLLs and soil lead level proximity to old city cores.

Genetic Variation

One of the reviewed studies focused on the genetic variation of childhood lead poisoning problems (Miranda 2005). Since other studies found a significant relationship between childhood lead poisoning and African-American populations, the authors focused on genetic variation of the problem. The study used previously developed data of children BLLs by Miranda et al. (2002), which geocoded children cases at the tax parcel level to get the construction year of house units from tax assessor data. The study also considers the occupancy status, which was also gathered from tax parcels. The authors note that the spatial autocorrelation problems were minimized by assigning individual year of construction from tax parcels. The ANOVA comparison of models with and without spatial autocorrelation also corroborated the nonexistence of spatial autocorrelation. Since some of the information pertaining to construction years is missing in the tax parcel data set, some cases lacked this information. In those cases, the study assigned the construction year from the nearby parcels. Some studies in the literature indicate that the relationship between high BLLs and African-American populations might be because of low calcium intake in the population. According to this study, however, the relationship between high BLLs and African-American populations might be more related to genetic polymorphisms (i.e., the simultaneous occurrence in the same locality of two or more discontinuous forms in such proportions that the rarest of them cannot be maintained by recurrent mutation).

Political Ecology

Hanchette's (2008) study focused on the political ecology aspect of childhood lead toxicity. The author used Moran's I (1950) and LISA statistics (Anselin 1995) to investigate the spatial distribution of lead poisoning prevalence at the county level in North Carolina. They use 10-year children BLL data from 1995 to 2004. In the study, the data findings show that there is a significant cluster of high BLL rates in eastern North Carolina. The author indicated that these clusters of high rates show persistent health disparities in the region. Hanchette claims that the health disparities in eastern North Carolina results from large-scale socioeconomic and cultural processes rather than neighborhood characteristics such as poverty and old housing. The study found that the Appalachia (western North Carolina) region displayed low rates of lead poisoning even though the region had high poverty rates. Another major finding is that high rates of lead poisoning clusters correspond with African-American populations only in eastern North Carolina. Unlike this region, southern North Carolina does not have high rates of lead poisoning despite high concentration of African-American populations. The author suggests that the convergence of poverty, older housing, and the large rural African-American population can be explained by the long history of tenant farming. According to Hanchette, this transition from an agricultural state to a mixed economy led to changes in socioeconomic characteristics of the eastern region of North Carolina.

Conclusion

GIS use in lead studies reveals greater detail about the magnitude of lead poisoning within populations. Surveillance and screening practices have assigned considerable amount of importance to targeting "at-risk" populations. This issue can be further resolved through the implementation of GIS in health departments. GIS-integrated screening could eliminate spatial bias due to disparities in reporting. Future studies are needed to fill this gap and attempt to improve the use of address geocoding in BLL data collection.

Future lead poisoning studies should also be concerned with data aggregation and the choice of geographical analysis. Studies show that finer geographic units such as census block group levels explain lead poisoning problems better, and hence some high levels of data aggregation (such as ZIP codes or census tracts) may not explain the distribution in the population (Krieger et al. 2003; Kaplowitz et al. 2010). Moreover, longitudinal lead studies are subject to possible errors as a result of change in census boundaries over time. In the latter case, very few studies examined the use of GIS and associated techniques to preserve confidentiality during the process of dissemination of screened children data and the resultant high-risk areas (Miranda 2011). Future studies focusing on environmental lead sources need to factor in intervention measures such as abatement efforts that may have taken place. By factoring in housing abatement efforts we can eliminate erroneous data and misinterpretations.

The environmental studies in this review also indicate a correlation between BLLs and African-American populations. However, very few studies investigated the individual characteristics of children (Miranda 2005). The history of socioeconomic and cultural processes could also be important factors to identify risk areas (Hanchette 2008). More GIS-based studies need to be conducted to investigate socioeconomic and cultural factors that focus on the development of an increasing awareness of the intricacies of lead poisoning and its effects on children and their neighborhoods.

References

Anselin, L. (1995). Local Indicators of Spatial Association—LISA. *Geographical Analysis* 27(2): 93–115. doi:10.1111/j.1538-4632.1995.tb00338.x.

Bocco, G. and Sánchez, R. (1997). Identifying potential impact of lead contamination using a geographic information system. *Environmental Management* 21(1): 133–138. doi:10.1007/s002679900012.

Bochynska, K. (2014). Facts and firsts of lead. Global lead advice & support service. http://www.lead.org.au/fs/fst29.html. Accessed May 16, 2014.

Brown, M.J. and Margolis, S. (2012). Lead in drinking water and human blood lead levels in the United States. *Morbidity and Mortality Weekly Report* Atlanta, GA: Center for Disease Control and Prevention. http://www.cdc.gov/mmwr/preview/mmwrhtml/su6104a1.htm.

Cayo, M.R. and Talbot, T.O. (2003). Positional error in automated geocoding of residential addresses. *International Journal of Health Geographics* 2: 10.

CDC. (1997). Update: Blood lead levels—United States, 1991–1994. *Morbidity and Mortality Weekly Report* 46(7): 141.

CDC. (2005). Blood lead levels—United States, 1999–2002. *Morbidity and Mortality Weekly Report* 54(20): 513–516.

CDC. (2014). Lead screening and prevalence of blood lead levels in children aged 1–2 years—Child blood lead surveillance system, United States, 2002–2010 and National Health and Nutrition Examination Survey, United States, 1999–2010. *Morbidity and Mortality Weekly Report* 63(02): 36–42.

Chisolm, J., Julian, E., Mellits, D., Keil, J.E., and Barrett, M.B. (1974). Variations in hematologic responses to increased lead absorption in young children. *Environmental Health Perspectives* 7: 7–12.

Committee on Environmental Health. (1998). Screening for elevated blood lead levels. *Pediatrics* 101(6): 1072–1078.

Cooper, M.H. (2014). Lead poisoning. CQ researcher by CQ press. http://library.cqpress.com/cqresearcher/cqresrre1992061900. Accessed February 20, 2014.

Cui, Y. (2013). A systematic approach to evaluate and validate the spatial accuracy of farmers market locations using multi-geocoding services. *Applied Geography* 41: 87–95.

Duncan, D.T., Castro, M.C., Blossom, J.C., Bennett, G.G., and Gortmaker, S.L. (2011). Evaluation of the positional difference between two common geocoding methods. *Geospatial Health* 5: 2.

Edwards, M. (2008). Lead poisoning: A public health issue. *Primary Health Care* 18(3): 18.

Edwards, M. (2014). Fetal death and reduced birth rates associated with exposure to lead-contaminated drinking water. *Environmental Science & Technology* 48(1): 739–746.

Farfel, M.R., Orlova, A.O., Lees, P.S.J., Rohde, C., Ashley, P.J., and Chisolm, J.J. (2003). A study of urban housing demolitions as sources of lead in ambient dust: Demolition practices and exterior dust fall. *Environmental Health Perspectives* 111(9): 1228–1234.

Goldberg, D.W. and Cockburn, M.G. (2012). The effect of administrative boundaries and geocoding error on cancer rates in California. *Spatial and Spatio-Temporal Epidemiology* 3: 1, 39–54.

Goldberg, D.W., Wilson, J.P., Knoblock, C.A., Ritz, B., and Cockburn, M.G. (2008). An effective and efficient approach for manually improving geocoded data. *International Journal of Health Geographics* 7(1): 60.

Gonzalez, E.J., Pham, P.G., Ericson, J.E., and Baker, D.B. (2002). Tijuana childhood lead risk assessment revisited: Validating a GIS model with environmental data. *Environmental Management* 29(4): 559–565.

Graber, L.K., Asher, D., Anandaraja, N., Bopp, R.F., Merrill, K., Cullen, M.R., Luboga, S., and Trasande, L. (2010). Childhood lead exposure after the phase out of leaded gasoline: An ecological study of school-age children in Kampala, Uganda. *Environmental Health Perspectives* 118(6): 884–889.

Graff, J.C., Murphy, L., Ekvall, S., and Gagnon, M. (2006). In-home toxic chemical exposures and children with intellectual and developmental disabilities. *Pediatric Nursing* 32(6): 596–603.

Gregorio, D.I., Cromley, E., Mrozinski, R., and Walsh, S.J. (1999). Subject loss in spatial analysis of breast cancer. *Health and Place* 5: 173–177.

Griffith, D.A., Millones, M., Vincent, M., Johnson, D.L., and Hunt, A. (2007). Impacts of positional error on spatial regression analysis: A case study of address locations in Syracuse, New York. *Transactions in GIS* 11(5): 655–679.

Griffith, D.A., Philip, G.D., Wheeler, D.C., and Johnson, D.L. (1998). A tale of two swaths: Urban childhood blood-lead levels across Syracuse, New York. *Annals of the Association of American Geographers* 88(4): 640–665.

Guthe, W.G., Tucker, R.K., Murphy, E.A., England, R., Stevenson, E., and Luckhardt, J.C. (1992). Reassessment of lead exposure in New Jersey using GIS technology. *Environmental Research* 59(2): 318–325.

Haley, V.B. and Talbot, T.O. (2004). Geographic analysis of blood lead levels in New York State children born 1994–1997. *Environmental Health Perspectives* 112(15): 1577–1582.

Hanchette, C.L. (2008). The political ecology of lead poisoning in eastern North Carolina. *Health and Place* 14(2): 209–216.

Hay, G., Kypri, K., Whigham, P., and Langley, J. (2009). Potential biases due to geocoding error in spatial analyses of official data. *Health and Place* 15: 2, 562–567.

Healthy People 2020 Objectives. (2014). Emerging Issues in Environmental Health: Blood Lead Levels. http://healthypeople.gov/. http://healthypeople.gov/2020/topicsobjectives2020/overview.aspx?topicid=12. Accessed February 24, 2014.

Jacquez, G.M. (2010). Space-time intelligence system software for the analysis of complex systems. In M.M. Fischer and A. Getis (Eds.), *Handbook of Applied Spatial Analysis*. Berlin, Germany: Springer, pp. 113–124. http://link.springer.com/chapter/10.1007/978-3-642-03647-7_7.

Joseph, C.L.M., Havstad, S., Ownby, D.R., Peterson, E.L., Maliarik, M., McCabe, M.J., Barone, C., and Johnson, C.C. (2005). Blood lead level and risk of asthma. *Environmental Health Perspectives* 113(7): 900–904. doi:10.1289/ehp.7453.

Kaplowitz, S.A., Perlstadt, H., and Post, L.A. (2010). Comparing lead poisoning risk assessment methods: Census block group characteristics vs. zip codes as predictors. *Public Health Reports* 125(2): 234–245.

Kim, D., Overstreet, M.A., Galeano, A.H., and Miranda, M.L. (2008). A framework for widespread replication of a highly spatially resolved childhood lead exposure risk model. *Environmental Health Perspectives* 116(12): 1735–1739.

Krieger, N., Chen, J., Waterman, P., Soobader, M., Subramanian, S., and Carson, R. (2003). Choosing area based socioeconomic measures to monitor social inequalities in low birth weight and childhood lead poisoning: The Public Health Disparities Geocoding Project (US). *Journal of Epidemiology and Community Health* 57(3): 186–199.

Laidlaw, M.A.S. and Filippelli, G.M. (2008). Resuspension of urban soils as a persistent source of lead poisoning in children: A review and new directions. *Applied Geochemistry* 23(8): 2021–2039.

Laidlaw, M.A.S., Mielke, H.W., Filippelli, G.M., Johnson, D.L., and Gonzales, C.R. (2005). Seasonality and children's blood lead levels: Developing a predictive model using climatic variables and blood lead data from Indianapolis, Indiana, Syracuse, New York, and New Orleans, Louisiana (USA). *Environmental Health Perspectives* 113(6): 793–800.

Litaker, D., Kippes, C.M., Gallagher, T.E., and O'Connor, M.E. (2000). Targeting lead screening: The Ohio lead risk score. *Pediatrics* 106(5): E69.

Lo, Y.C., Dooyema, C.A., Neri, A. et al. (2012). Childhood lead poisoning associated with gold ore processing: A village-level investigation–Zamfara State, Nigeria, October-November 2010. *Environmental Health Perspectives* 120(10): 1450–1455.

Lovasi, G.S., Weiss, J.C., Hoskins, R., Whitsel, E.A., Rice, K., Erickson, C.F., and Psaty, B.M. (2007). Comparing a single-stage geocoding method to a multi-stage geocoding method; how much and where do they disagree? *International Journal of Health Geographics* 6: 12.

Lutz, J., Jorgensen, D., Hall, S., and Julian, J. (1998). Get the lead out: A regional approach to healthcare and beyond. *Geo Info Systems* 8(7): 26–30.

McDonald, J.A. and Potter, N.U. (1996). Lead's legacy? Early and late mortality of 454 lead-poisoned children. *Archives of Environmental Health* 51(2): 116–121.

Mielke, H.W., Dugas, D., Mielke Jr., P.W., Smith, K.S., Smith, S.L., and Gonzales, C.R. (1997). Associations between soil lead and childhood blood lead in urban New Orleans and rural Lafourche Parish of Louisiana. *Environmental Health Perspectives* 105(9): 950–954.

Mielke, H.W., Gonzales, C.R., and Mielke Jr., P.W. (2011a). The continuing impact of lead dust on children's blood lead: Comparison of public and private properties in New Orleans. *Environmental Research* 111(8): 1164–1172.

Mielke, H.W., Gonzales, C.R., Powell, E.T., and Mielke, P.W. (2013). Environmental and health disparities in residential communities of New Orleans: The need for soil lead intervention to advance primary prevention. *Environment International* 51(January): 73–81.

Mielke, H.W., Laidlaw, M.A.S., and Gonzales, C. (2010). Lead (Pb) legacy from vehicle traffic in eight California urbanized areas: Continuing influence of lead dust on children's health. *Science of the Total Environment* 408(19): 3965–3975.

Mielke, H.W., Laidlaw, M.A.S., and Gonzales, C.R. (2011b). Estimation of leaded (Pb) gasoline's continuing material and health impacts on 90 US urbanized areas. *Environment International* 37(1): 248–257.

Miranda, M.L., Anthopolos, R., and Hastings, D. (2011). A geospatial analysis of the effects of aviation gasoline on childhood blood lead levels. *Environmental Health Perspectives* 119(10): 1513–1516.

Miranda, M.L. and Dolinoy, D.C. (2005). Using GIS-based approaches to support research on neurotoxicants and other children's environmental health threats. *NeuroToxicology* 26(2): 223–228.

Miranda, M.L., Dolinoy, D.C., and Overstreet, M.A. (2002). Mapping for prevention: GIS models for directing childhood lead poisoning prevention programs. *Environmental Health Perspectives* 110(9): 947–953.

Miranda, M.L., Kim, D., Hull, A.P., Paul, C.J., and Overstreet, M.A. (2007). Changes in blood lead levels associated with use of chloramines in water treatment systems. *Environmental Health Perspectives* 115(2): 221–225.

Moran, P.A.P. (1950). Notes on continuous stochastic phenomena. *Biometrika* 37(1/2): 17–23.

National Center for Environmental Health. (2014). CDC—Lead—Screening document 97. http://www.cdc.gov/nceh/lead/publications/screening.htm. Accessed May 18, 2014.

Oliver, M.N., Matthews, K.A., Siadaty, M., Hauck, F.R., and Pickle, L.W. (2005). Geographic bias related to geocoding in epidemiologic studies. *International Journal of Health Geographics* 4: 29.

Oyana, T.J. and Margai, F.M. (2007). Geographic analysis of health risks of pediatric lead exposure: A golden opportunity to promote healthy neighborhoods. *Archives of Environmental and Occupational Health* 62(2): 93–104.

Oyana, T.J. and Margai, F.M. (2010). Spatial patterns and health disparities in pediatric lead exposure in Chicago: Characteristics and profiles of high-risk neighborhoods. *The Professional Geographer* 62(1): 46–65.

Preventing Lead Poisoning in Young Children: A Statement by the Centers for Disease Control, March 1975. (2014). http://www.cdc.gov/nceh/lead/publications/plpyc1975.pdf. Accessed February 21, 2014.

Preventing Lead Poisoning in Young Children: A Statement by the Centers for Disease Control, January 1985. (2014). http://www.cdc.gov/nceh/lead/publications/plpyc1985.pdf. Accessed February 21, 2014.

Preventing Lead Poisoning in Young Children: A Statement by the Centers for Disease Control, October 1991. (2014). http://www.cdc.gov/nceh/lead/Publications/books/plpyc/contents.htm. Accessed February 21, 2014.

Preventing Lead Poisoning in Young Children: A Statement by the Centers for Disease Control and Prevention, August 2005. (2014). http://www.cdc.gov/nceh/lead/publications/prevleadpoisoning.pdf. Accessed February 21, 2014.

Public Health and Welfare. (2014). Prohibition on use of lead pipes, solder, and flux. U.S.C. http://www.gpo.gov/fdsys/granule/USCODE-1999-title42/USCODE-1999-title42-chap6A-subchapXII-partB-sec300g-6/content-detail.html. Accessed February 25, 2015.

Reed, A.J. (1972). Lead poisoning—Silent epidemic and social crime. *The American Journal of Nursing* 72(12): 2180–2184.

Reissman, D.B., Staley, F., Curtis, G.B., and Kaufmann, R.B. (2001). Use of geographic information system technology to aid Health Department decision making about childhood lead poisoning prevention activities. *Environmental Health Perspectives* 109(1): 89–94.

Roberts, J.R., Hulsey, T.C., Curtis, G.B., and Reigart, J.R. (2003). Using geographic information systems to assess risk for elevated blood lead levels in children. *Public Health Reports* 118(3): 221–229.

Sargent, J.D., Bailey, A., Simon, P., Blake, M., and Dalton, M.A. (1997). Census tract analysis of lead exposure in Rhode Island children. *Environmental Research* 74(2): 159–168.

Summary of the HIPAA Privacy Rule. (2014). HIPAA Privacy Rule. http://www.hhs.gov/ocr/privacy/hipaa/understanding/summary/index.html. Accessed May 17, 2015.

Summary of the HIPAA Security Rule. (2014). HIPAA Security Rule. http://www.hhs.gov/ocr/privacy/hipaa/understanding/srsummary.html. Accessed May 17, 2015.

Trasande, L. and Liu, Y. (2011). Reducing the staggering costs of environmental disease in children, estimated at $76.6 billion in 2008. *Health Affairs* 30(5): 863–870.

United States Environmental Protection Agency. (2014). Basic information—fuels and fuel additives. Washington, DC: U.S. EPA. http://www.epa.gov/otaq/fuels/basicinfo.htm. Accessed February 25, 2015.

U.S. Consumer Product Safety Commission. (2014). Ban of lead-containing paint and certain consumer products bearing lead-containing paint. http://www.ecfr. gov/cgibin/retrieveECFR?gp=1&SID=1873431f79f14a0e5a89f773986741c1&ty= HTML&h=L&r=PART&n=16y2.0.1.2.66#16:2.0.1.2.66.0.1.1. Accessed February 25, 2015.

Vach, W. and Blettner, M. (1991). Biased estimation of the odds ratio in case-control studies due to the use of ad hoc methods of correcting for missing values for confounding variables. *American Journal of Epidemiology* 134: 895–907.

Vaidyanathan, A., Staley, F., Shire, J., Muthukumar, S., Kennedy, C., Meyer, P.A., and Brown, M.J. (2009). Screening for lead poisoning: A geospatial approach to determine testing of children in at-risk neighborhoods. *The Journal of Pediatrics* 154(3): 409–414.

Ward, M.H., Nuckols, J.R., Giglierano, J., Bonner, M.R., Wolter, C., Airola, M., Mix, W., Colt, J.S., and Hartge, P. (2005). Positional accuracy of two methods of geocoding. *Journal of Epidemiology* 16(4): 542–547.

Wey, C.L., Griesse, J., Kightlinger, L., and Wimberly, M.C. (2009). Geographic variability in geocoding success for West Nile virus cases in South Dakota. *Health and Place* 15(4): 1108–1114.

Whitsel, E.A., Quibrera, P.M., Smith, R.L., Catellier, D.J., Liao, D., Henley, A.C., and Heiss, G. (2006). Accuracy of commercial geocoding: Assessment and implications. *Epidemiologic Perspectives and Innovations* 3: 8.

WHO. (2014). Childhood lead poisoning. Geneva, Switzerland: WHO. http://www. who.int/ceh/publications/childhoodpoisoning/en/. Accessed May 16, 2015.

WHO. (2015). Childhood lead poisoning. Geneva, Switzerland: WHO. http://www. who.int/ceh/publications/childhoodpoisoning/en/. Accessed March 21, 2015.

Wu, J., Funk, T.H., Lurmann, F.W., and Winer, A.M. (2005). Improving spatial accuracy of roadway networks and geocoded addresses. *Transactions in GIS* 9(4): 585–660.

Yasnoff, W.A. and Sondik, E.J. (1999). Geographic information systems (GIS) in public health practice in the new millennium. *Journal of Public Health Management and Practice: JPHMP* 5(4): ix–xii.

Zandbergen, P.A. (2008). A comparison of address point parcel and street geocoding techniques. *Computer Environment and Urban Systems* 32: 3, 214–232.

6

GIS's Applications in Health-Care Access

The amount and type of health-care services populations receive depend greatly on where they live, the capacity of the health-care system in their area, and the methods practiced by local providers. Different populations need particular services, and they need to be located within reasonable distances of the services they need. Knowing the demographics of an area, where health problems are prevalent, the demand for any particular service, and standards of care may help health-care organizations to decide how to allocate finite health-care resources and where to build new facilities and design population-based interventions.

Measuring the spatial accessibility (i.e., based on travel times between residents and physicians or areas within a certain travel time of each other), aggregating various sociodemographic variables to administrative boundaries, calculating geographic centroids and population-weighted centroids, geocoding physician locations, simple overlays, clustering of utilization are fundamental geographic information systems (GIS) functions applied in health-care resource planning. Computing accessibility measures in GIS using network distances and more precise aggregation methods is no longer a daunting task. The time required for the computation of numerous network distances or aggregating voluminous data to high-level census boundaries is no longer a limitation. Most practicing physicians and policy makers have access to Google Maps, in-house GIS, and/or online GIS tools (e.g., ArcGIS. com map viewer published by ESRI, ArcGIS users, or other data providers) and interactive maps with the capability to understand the neighborhoods of their patients and available resources nearby that help treatment and intervention efforts (Berke 2010).

More sophisticated work has been done with GIS to define catchment areas (i.e., the road distance to the nearest care center) of the health-care centers (Radke and Mu 2000): evaluation of accessibility with location-allocation modeling (Barnes and Peck 1994; Lang 2000), evaluation of optimal routes for health service delivery, development of space–time models for effective allocation of resources (Keenen et al. 2007), and site selection of health-care facilities.

Most frequently, GIS has been used in the assessment of accessibility and utilization of health-care services (Cromley and McLafferty 2002). As studies at coarser levels of aggregation can miss important local excesses and underlying mechanisms at a finer scale, GIS and recent advances in computer technology made small-area health-care variation studies more feasible (Martin et al. 1996; Stukel et al. 2005, 2007; Anthony et al. 2009; Magner et al. 2009;

Mulley 2009; Amico et al. 2010; Lucas et al. 2010; Makarov et al. 2010; Matlock et al. 2010; O'Hare et al. 2010; Welch et al. 2011). Beginning in the 1990s, with the recognition of GIS's potential to examine data in a spatial context, the health-care delivery system in the United States channeled research efforts to effectively identify areas with primary care shortages and target resources to benefit the underserved populations (Lee 1991; General Accounting Office 1995). Until recently, there has also been increasing interest in using GIS for the evaluation of the effectiveness of intervention measures as a new trend in the geography of health (Renger et al. 2002; Caley 2004; Jankowski and Nyerges 2001). GIS now plays a critical role in determining where and when to intervene, increasing accessibility of health service, finding more cost-effective delivery modes, improving the quality of care, and preserving patient confidentiality while satisfying the needs of the research community for data accessibility. This chapter describes GIS's role in determining disparities in health-care access, the diverse health data sets used in small-area health-care variation studies, and data limitations. The final sections examine spatial issues in rural and regional disparities, health profession distribution, and evaluation of the effectiveness of intervention measures.

GIS's Role in Determining Disparities in Health-Care Access

GIS gives detailed and compelling answers to the questions health policy makers ask, Who is underserved? Where? Why? These questions are equally important in health disparities research. Access to health care varies across space as well as among certain population groups (Wang and Luo 2005; Agency for Healthcare Research and Quality National Healthcare Disparities Report 2008). The Institute of Medicine has stated that there could be systemic disparities in health-care access (e.g., access to emergency departments [EDs]) among traditionally vulnerable patients, including those of racial/ethnic minorities, foreign-born, the economically disadvantaged (i.e., low income, unemployed), the elderly, and rural populations (Committee on the Future of Emergency Care in the United States Health System 2007). For example, in the United States, African-American and Latinos are less likely to receive care that is responsive to complex medical and social needs and hence rely on ED care when conditions are exacerbated or symptoms escalate than whites are, even after controlling for sociodemographic differences (Zuvekas and Weinick 1999; Weinick et al. 2000; Walls et al. 2002; Ludwig et al. 2009). Minorities also are less likely to receive preventive care and their rates of preventable hospitalizations are substantially higher than those of whites (Gaskin and Hoffman 2000; Hargraves and Hadley 2003). In order to address rehospitalizations, the Affordable Care Act payment reform has imposed penalties to hospitals with excess readmissions beginning in

2012 (Billings et al. 2013). Decision-makers in public health are very much interested in identifying subpopulations who more likely to overutilize health services (i.e., high repeat utilizers). Targeting resources toward high repeat utilizers has the potential for not only improving quality of care but also eliminating costly health-care expenditures (e.g., inpatient rehospitalization). GIS has been used to identify geographic variation in rates of readmission among health systems (Bernheim et al. 2010; Bernheim et al. 2013; Maheswaran and Maheswaran 1997; Majeed et al. 2002).

The extent to which health services provide access to care for underserved and uninsured populations merits attention in both health-care access and health disparities research. GIS serves as an entry point to the health-care system to identify medically underserved areas (MUAs) (i.e., disadvantaged neighborhoods) and individuals without access to a primary care provider, those without health insurance, or those unable to get the care they need in a timely fashion. Research has shown the lower geographic accessibility of safety net health-care providers (i.e., health system provides a significant level of care to low-income, uninsured, and vulnerable populations) was associated with higher levels of unmet medical needs among uninsured individuals and individuals without a regular provider (Hadley and Cunningham 2004). People without insurance or a regular provider may have limited options for receiving care, and thus, people may be more willing to travel longer distances to receive medical care.

In the United States, EDs are increasingly being used by individuals without insurance and a primary care doctor as a provider of last resort for nonurgent care (Wilner 1977; Grumbach et al. 1993; Suruda et al. 2005; Rust et al. 2008). Underserved and uninsured populations who seek ED care for needs that are best addressed through preventative measures and by primary care providers could be channeled to the appropriate care in the right setting while reducing crowding and waiting times and the financial burden in the EDs (Derlet et al. 2001; Lambe et al. 2002). Primary care facilitates the building of doctor–patient relationships and access to treatment adherence, follow-up, and health education while decreasing inappropriate use of EDs, hospitalization rates, and tests requested; this would all subsequently lower associated costs (Carret et al. 2007). Research that utilize GIS examine geographic access deterioration to ED (Shen and Hsia 2010) and the association between ED utilization and geographic variables related to the health-care environment (American Hospital Association 2001; Rafalski and Zun 2004; Weber et al. 2005). GIS maps show utilization patterns, temporal change in population access and driving time to ED, high concentration areas and populations of ED usage, the "hot spots," in order to better understand the primary care needs of the community and the costs of using the ED (Gawandi 2011). The interpretation of these maps helps with interventions how to transform health care away from ED use and instead reinvest in these "hot spot" communities.

Opening retail clinics in chain stores is a private sector alternative for the uninsured to avoid ED use at traditional care facilities for unurgent visits such as acute illnesses and immunizations (Kershaw 2007, 2008; Mehrotra et al. 2008; Convenient Care Association 2008). Lower prices with greater cost transparency and convenient hours with lack of scheduled appointments are primary reasons why retail clinics are preferred by the uninsured (Tu and Cohen 2008), though stores with retail clinics tend to be located in more advantaged neighborhoods with more white residents, fewer blacks and Hispanic residents, higher rates of home ownership, higher median incomes, and lower rates of poverty compared with stores without retail clinics (Pollack and Armstrong 2009). Despite retail clinics are currently located in more advantaged neighborhoods (Pollack and Armstrong 2009), surveys resulted that approximately a quarter of retail clinics users are uninsured and without a regular provider (Hadley and Cunningham 2004; Retail-Based Clinic Policy Work Group, AAP 2006). Rethinking the geographic distributions and future site selection for retail clinics (i.e., distribution of chain stores with retail clinics) is an application of GIS both public and private policy makers could benefit from, in order to improve access for uninsured populations.

GIS is also used in exploring issues such as utilization and supply of health-care services (Trendwatch 2000; DAP 2015). Health-care utilization patterns and supply of health-care services could differ from community to community, from region to region, from country to country. Uninsured and underserved populations vary from place to place and their needs differ accordingly. Every country has a social safety net, a collection of services provided by the state or other institutions, for people unable to support themselves. These could include but not limited to welfare, unemployment benefit, universal health care, homeless shelters, and sometimes subsidized services such as public transport. GIS has potential to integrate data from various systems these services use and work with the data geographically. This allows getting more from the data by using it in new ways and being able to evaluate the effectiveness of the intervention measures.

America's safety net is often thought of as one big network of similar providers with common patient populations and funding streams. A geographical analysis of hospital data reveals that service delivery methods differ by region (American Hospital Association 2000, 2001; Institute of Medicine 2000). Some communities serve uninsured and underinsured populations with public hospitals and health systems bearing the primary responsibility and others spread care of the poor among many area hospitals. Houston, with a large Hispanic population of undocumented immigrants, is an example of an area that relies on a public hospital. The only public hospital in the area provides 36% of the area's Medicaid and uncompensated care. In addition, five hospitals combine to provide 70% of care to the uninsured. The uninsured patients in Houston may need to travel to a centralized location. In contrast, the Detroit area distributes indigent (i.e., poor) care among a larger group of hospitals. Hospitals can spread out the financial burden of serving the uninsured among a greater number of institutions and uninsured patients may not need to travel as far to receive care because hospitals providing charity care are located in geographically diverse areas of the metropolitan statistical area.

San Francisco, which has a large Asian community, is a metropolitan area that has high population densities and high absolute numbers of people without insurance. These communities have designed and implemented very different methods of providing for the needy in their communities. For example, transport to community-based services, mobile clinics, and prevention and wellness programs that help low-income patients, patients with high medical costs, and uninsured patients (Sutterhealth 2015).

GIS is used in public health reporting of health-care disparities (U.S. Department of Health and Human Services 2000a), evaluation of nationally representative data sources whether they produce consistent estimates of disparities (Bilheimer and Sisk 2008; Bilheimer and Klein 2010), and indirect methods of estimation to identify racial and ethnic disparities in health-care access (Tufte 2001). In the United States, the Patient Protection and Affordable Care Act (P.L. 111-48, 2009) requires any federally conducted or supported health-care or public health program, activity, or survey to collect and report data on race, ethnicity, sex, primary language, and disability status, to the extent feasible, to collect these data at the smallest geographic level possible, and to have sufficient data to generate reliable estimates for subgroups of these populations. With this Act, most attention has focused on data limitations relating to race and ethnicity and identifying disparities in subpopulations. Some health plans seek to identify racial and ethnic disparities in health-care access and quality by using indirect methods of estimation, as direct collection of this information has proven difficult (Higgins and Taylor 2009). Indirect methods use different combinations of surname analysis, geocoding, and Bayesian techniques with effectiveness varying by subgroup (Fiscella and Fremont 2006; Elliott et al. 2008a, 2008b). GIS is used to derive indirect race and ethnicity information for health plan members (Tufte 2001). For example, GIS's geocoding tool determines the physical location of a health plan enrollee's address that has the benefit of allowing members to be tied to census data such as education levels, language proficiency, and income (Krieger 1992; Gornick et al. 2003). The addition of Bayesian techniques to surname analysis and geocoding allows estimation of the probability that an individual is in a particular group, and it appears to improve estimates of race and ethnicity relative to other indirect methods. This approach is useful for determining the racial and ethnic composition of a health plan's enrollees and identifying disparities among groups (Porter 2010).

Data for Health Services Resources and Planning

There are technological advances in communication where information about health services could be distributed through the mass media, social networking, and the Internet (Green and Himelstein 1998), health GIS pioneers have encouraged a growing number of state and local agencies to publish their health statistics, using GIS to query rapidly and interactively large volumes of health data in aggregated format at different spatial scales and population levels and to produce meaningful results displayed as tables, graphs, and maps. Small-area studies, in which health-care data are analyzed with much finer geographical resolution than has hitherto been possible, represent an important approach to determine how much of the geographic variation in rates among populations is associated with disparities. The Dartmouth Atlas of Health Care Project is one of the pioneer projects that produces maps showing variations in how health resources are distributed in the United States. The project provides Medicare data using geographic boundaries aligned with health-care delivery systems rather than political boundaries such as states and counties. In the United States, however, there is no administrative data set covering the entire population, although increasing adoption of electronic health records provides future potential for population-wide data. Although Medicare provides health care for an estimated 96% of the elderly, age 65 years and older, there is no comparable source for those under 65. Medicare data are a comprehensive data set about health-care utilization and payments for procedures, services, and prescription drugs provided to Medicare beneficiaries by specific inpatient and outpatient hospitals, physicians, and other suppliers (Vinig and McBean 2001). The Dartmouth project aggregates Medicare beneficiaries' data to regions that characterize health service areas that often offer an important resource to further examine patterns of utilization and the possible interactions with population health outcomes and health-care expenditures. These data however reflect the experience of Medicare patients living in the region, regardless of where the care was actually delivered. Dartmouth Atlas of Health Care Project arranges data in three hierarchical levels, from zip codes to hospital service areas (HSAs) and hospital referral regions (HRRs). An HSA is a collection of ZIP codes whose residents receive most of their hospitalizations from the hospitals in that area. ZIP codes are assigned to the hospital area where the greatest proportion of their Medicare residents was hospitalized. HRRs were defined by determining where patients were referred for tertiary medical care (i.e., major cardiovascular surgical procedures and for neurosurgery) that generally requires the services of a major referral center. Each HSA was examined to determine where most of its residents went for these services. For example, these data can help develop treatment profiles that describe how often certain treatments or procedures are performed in comparable populations. In some regions, doctors perform more bypass grafts, an expensive cardiac surgery procedure, but in other areas, prefer balloon angioplasty, which costs less. GIS gives the capability to query for data on these procedures (per one thousand patients) for each HRR and can give an idea of how costs vary in the different markets and how it compares with national averages and evaluate the availability of resources (e.g., amount of cardiovascular specialists).

Public health researchers and health planners can also work with a variety of other data geographically, such as MUAs, health professional shortage areas (HPSAs), and primary care service areas (PCSAs). In the United States, research in the health-care delivery system started in 1976 by developing MUA system to identify areas eligible for federally funded community health centers. Followed up in 1978, HPSAs were designated to direct replacement of National Health Service Corps employees to counties with shortage of physicians (Meade and Erickson 2000). Now, the methods to designate these areas are improved and implemented using GIS. U.S. Department of Health and Human Services integrates spatial (i.e., spatial accessibility) and nonspatial (i.e., socioeconomic) factors to identify these areas and population groups. MUA is based on the ratio of primary care physicians per 1000 population, infant mortality ratio, percentage of the population with incomes below the poverty level, and the percentage of the population 65 years and older. MUA is designated for census tracts in metropolitan areas and counties in nonmetropolitan areas. HPSAs are designated for primary health care, mental health services, and dental services. Figure 6.1 shows HPSAs in the southeastern United States. HPSAs can be located within major cities

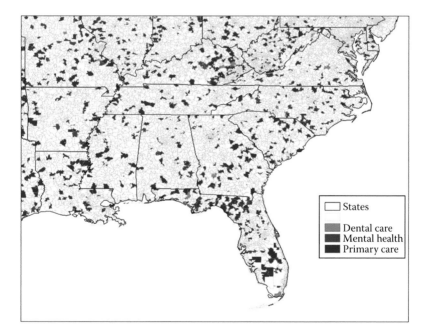

FIGURE 6.1
Health professional shortage areas (HPSAs) in the southeastern United States. HPSA data are downloaded from Health Resources and Services Administration Data Warehouse. (From U.S. Department of Health and Human Services, Health resources and services administration, http://datawarehouse.hrsa.gov/tools/dataportal.aspx, 2000.)

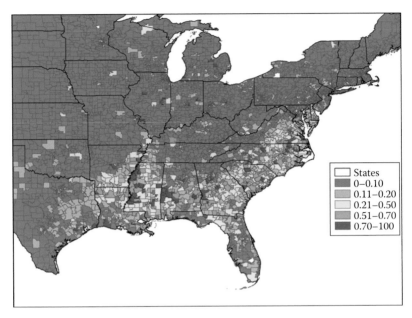

FIGURE 6.2
Primary care service areas (PCSAs) with percent African-American population in the southeastern United States. PCSA data are downloaded from Health Resources and Services Administration Data Warehouse. (From U.S. Department of Health and Human Services, Health resources and services administration, http://datawarehouse.hrsa.gov/tools/dataportal.aspx, 2000.)

and near major hospitals and clinics. PCSAs reflect Medicare patient travel to primary care providers. Figure 6.2 shows PCSAs with percent African-American population in the southeastern United States.

The Health Resources and Services Administration (HRSA) Data Warehouse (HDW) integrates HRSA data with external sources, such as the U.S. Census Bureau, enabling users to download and transfer data into GIS and gather relevant and meaningful information about health-care programs (e.g., MUAs, PCSAs) and the populations they serve. HDW is used by HRSA, grantees, care providers, the public, and other audiences interested in HRSA's public health services and information (U.S. Department of Human Services 2000b). Since the issue of access to health services is in many respects a geographical one, spatial display of data is essential. HDW provides interactive maps to visualize health-care programs and populations they serve. The statistics in the data viewer are for looking at small-area health-care statistics to see how services are used by defined populations. Chronic Condition Data Warehouse (CCW) also provides population health and health-care data at the HRR, state, and county levels (CMS 2015). CCW presents prevalence and Medicare utilization and spending for the 17 chronic conditions. Their reports help compare geographic areas to national Medicare estimates.

In addition to countrywide health information systems, a web-based national source of proprietary hospital and health system data includes exclusive national survey data (e.g., AHA Annual Survey of Hospitals), proprietary AHA membership data, and hospital financial data from Medicare Cost Reports. Hospitals have a history of collecting race, ethnicity, and primary language data (AHA 2008). Special GIS-based studies involve hospital groups, linking these data to quality measures and health disparities (Weinick et al. 2007). For example, a Robert Wood Johnson Foundation initiative to reduce disparities in cardiac care required participating hospitals to systematically collect race, ethnicity, and language data and use the data to stratify quality measures. Financial data from the Annual Survey of Hospitals could be integrated into GIS and uncompensated care could be mapped. Financial data also are used to support sales, business development, planning, and marketing. GIS is instrumental exploring variations in treatments and costs in health-care markets. The use of GIS and its spatial analysis functions has enhanced the marketers' ability to identify and characterize the health system stakeholders (i.e., patients, doctors, pharmacies, and medical facilities) to achieve a more accurate and detailed representation of need of services or for a particular product. For example, hospitals use GIS to tailor its services to the needs of employers by locating specialized rehabilitation centers near them. Hospitals may soon be able to use GIS to create real-time maps that track the movement of their personnel and patients. By being able to accurately monitor the movement of employees and registered patients, hospital can guarantee safety and security, in sensitive departments like maternity ward (Lang 2000).

The National Health Care Surveys are designed to answer key questions related to the factors that influence the use of health-care resources; the quality of health care, including safety; and disparities in health-care services provided to population subgroups in the United States (Davern et al. 2008). In the case of national surveillance and annual snapshot reports, *Healthy People 2010* uses National Health Interview Survey (NHIS) data (Clarke et al. 2015) to monitor insurance coverage and access to a usual source of care and uses National Vital Statistics System data to monitor access to prenatal care (U.S. Department of Health and Human Services 2000a). Nationally representative population data sources are the Current Population Survey (CPS), the NHIS, and the Medical Expenditure Panel Survey (MEPS). The CPS collects extensive demographic data that complement and enhance labor market conditions in the nation overall, among many different population groups, in the states and in substate areas (U.S. Bureau of Labor Statistics 2006). NHIS data on a broad range of health topics are collected through personal household interviews. The U.S. Census Bureau has been the data collection agent for the NHIS. The MEPS is a set of large-scale surveys of families and individuals, their medical providers, and employers across the United States. MEPS is the most complete source of data on the cost and use of health care and health insurance coverage.

Physician Masterfile of the American Medical Association (AMA) is a good resource for defining HPSAs. The AMA Division of Survey and Data Resources is dedicated to effectively and accurately collect, analyze, and manage data within the Masterfile that serves as a primary resource for professional medical organizations (AMA 2015). Physicians are geocoded to zip code areas since many records only have "P.O. Box" addresses with zip codes but not street addresses. These data are integrated with census population data to designate HPSAs. Most of the population group HPSAs are low-income or minority groups (DHHS 2004). Spatial accessibility based on travel between residents and physicians is measured using GIS functions. Measurement methods of spatial accessibility are implemented in a GIS environment.

A variety of marketplace data are also collected, analyzed, and distributed with GIS by a private sector for efficient health delivery, such as physician prescription–writing data, collected by pharmaceutical companies. They look for areas where the physicians most likely to prescribe a drug and draw their sales territories accordingly and make more efficient routes. Manufacturers of medical equipment use GIS to find out which diseases and conditions are most common in an area so they know which of their products to market there. Insurance companies use GIS to make sure their members are within reasonable distances of the services they need (Lang 2000).

Data Limitations in Small-Area Health-Care Variation Studies

In the small-area analysis, there are small populations and small-area variation challenges to estimating rates at the level of local health-care delivery systems. There may be significant over dispersion due to unstable rates in small populations (high rates in small areas). Stratified analyses may be limited to areas with sufficient number of racial and ethnic minorities.

Interest is growing in the collection of more granular race and ethnicity data, moving beyond self-reporting from a limited list of race/ethnicity options to reporting self-identified race and/or ethnicity (Prewitt 2005; Hasnain-Wynia and Baker 2006). There are several reasons why this approach is considered. Self-report data come through unknown combinations (i.e., patients, family members, and other observations); cross-cultural biases in surveys may disguise health disparities or contribute to apparent differences (Hillenmaier et al. 2007; Leung et al. 2007; Carle 2009); people from different countries of origin may not self-identify themselves with any of the listed racial and ethnic groups based on the Office of Management and Budget categories (Executive Office of the President 1997). For example, more than 40% of Hispanics reported themselves as "some other race" in the Census 2000 (Grieco and Nassidy 2001).

Self-identified race/ethnicity data bring up other complex measurement issues. Computationally feasible algorithms have not yet been developed to address the potential heterogeneity of self-identified race/ethnicity responses. How to obtain the denominator data is another complex situation. The crude ratio of a health indicator presents the exact measure with the number of cases as the numerator and the total population as the denominator. This type of information is useful to inform on health and health access problems in a specific population. The adjusted measure takes into account the variations in race and ethnicity among populations for comparison between them. If health-care organizations succeed in collecting detailed race and ethnicity information on all their patients, they have the denominators. They can explore health disparities in their specific patient population. But, if they want to assess disparities in health-care access in their communities and make comparisons, then they need the same granular information for the local population as a whole. However, community-level rates might have to be based on aggregated data, because current data sources (e.g., census and the American Community Survey [ACS]) cannot handle resulting population estimates introducing errors. The ACS releases 1-, 3-, and 5-year estimates of ancestry information. Even combining 5 years of data, however, standard errors of estimates may be large for small racial, ethnic, and ancestry groups. Five-year estimates, moreover, will be slow to pick up rapid changes in population composition. The duration of data is of particular importance to small-area analyses, because they are often based on small populations with few cases and therefore need several years of data to give stability to any type of estimation.

Different data sources could produce substantially different estimates for the same population. Conclusions about health-care disparities may be influenced by the data source used. For example, Johnson et al. (2010) examined rates of uninsurance among the American Indian/Alaska Native (AIAN) population using three surveys: CPS, NHIS, and MEPS. Sociodemographic differences across surveys suggest that national samples produce differing estimates of the AIAN population (Johnson et al. 2010; Steele et al. 2008). AIAN all-year uninsurance rates varied across surveys (3%–23% for children and 18%–35% for adults). Measures of disparity also differed by survey. For all-year uninsurance, the unadjusted rate for AIAN children was 2.9 times higher than the rate for white children with the CPS, but there were no significant disparities with the NHIS or MEPS. Compared with white adults, AIAN adults had unadjusted rate ratios of 2.5 with the CPS and 2.2 with the NHIS or MEPS. With this research, Johnson et al. (2010) demonstrated how choice of data source impacts conclusions about health-care disparities.

Small-area health-care variation studies increasingly rely on GIS methods for integrating health indicator data across spatial and temporal scales such as census tracts and zip codes at multiple time periods. Health-care systems currently offer health indicator data including incidence, morbidity, mortality, and health service utilization (Vanasse et al. 2012; Mulholland et al. 1996).

Medical consultations and drug treatments indicator data are equally important in small-area analysis, because variation in health resource utilization could be driven not only by the characteristics of patients or populations but also differences in physician practice style and health resource supply (Wennberg 1984, 1987; Joines et al. 2003). There is also considerable evidence supporting the contribution of community socioeconomic characteristics for certain health conditions (Folland and Stano 1990; Billings et al. 1993; McMahon et al. 1993). Even though a more comprehensive set of health indicators data might better fit the needs of decision-makers to assist them in health resource allocation, linking data from various data sources gives rise to spatial uncertainties. An example of this situation is that utilization rates are available at the hospital level, where socioeconomic covariates come at census tract level. This type of spatial mismatch among different data sources has long been identified as a change of support problem in the literature (Gelfand et al. 2001). However, concrete statistical methodology has not yet been developed to handle the mismatches via computationally feasible algorithms.

Translating a hospital-based measure to health administrative boundaries has strong points and also limitations for population-based reporting. Where point data fail to provide a continuous coverage or the patterns become confusing to interpret, areal-based analysis is essential to bring order and understanding to the spatial distribution of a variable of interest. There is a substantial literature on the use of area measures in health research (Krieger et al. 2003a,b,c; Krieger 2005). Comparing the effects of using data aggregated to various geographic levels, generally, the conclusion has been that effects are detected more sensitively when data are linked to smaller (more detailed) geographic units. Certain variables may need to be derived from census information. For example, aggregate income or education measures at the zip code and census tract level from GIS are used as proxy for a patient's income or education (Krieger 1992, 2005; Gornick et al. 2003).

On the contrary, researchers should use caution in substituting aggregate for individual-level variables (e.g., hospital-based measure), because of potential biases and confounding problems (Geronimus et al. 1996). Analysts who base their conclusions about individual-level relationships on area-level analyses run the risk of committing the ecological fallacy. Moreover, the results of the analysis of area-level data may vary according to the number of areas used and their boundaries; this is referred to as the modifiable areal unit problem (MAUP). The impact of the choice of geographic unit or the MAUP problem (Openshaw 1984) could be assessed by analyzing the data using alternative units. This approach could be limited by lack of available data at that unit level and the instability of rate estimates based on small number problem.

Another important data limitation is the spatial data's natural tendency to cluster. The resultant spatial patterns could result from spatial dependence in the data. Spatial autocorrelation is the term used for spatial dependence

(Cliff and Ord 1981; Anselin 1992a,b). For example, positive spatial autocorrelation refers to tendency for clustering of nearby areas with similar attributes. Hospitalization rates in one county could influence rates in neighboring counties. There could be underlying mechanisms that cause this spatial effect. For example, interaction between doctors, feedback from specialist from referral centers, and continuing medical education activities could diffuse and disseminate to nearby counties (Joines et al. 2003). Competition between hospitals to meet patient expectations for appropriate care could induce similar patterns (Joines et al. 2003). When spatial autocorrelation in the data renders the usual multiple regression models (i.e., estimated by the method of ordinary least squares), studies of small-area variation should consider alternative spatial models that examine and control for spatial autocorrelation. For example, the spatial lag and spatial error models incorporate spatial autocorrelation in the data (Anselin 1992a,b). The spatial lag regression models reveal an underlying spatial process not accounted for by the explanatory variables in the regression models. Broader use of spatial lag regression modeling techniques in small-area health-care variation studies is suggested.

Issues of data reliability, validity, and accuracy are very important when using spatial databases provided by a commercial, government, or other providers. Seeking a metadata file that describes the health data resources that complies with the Federal Geographic Data Committee standards should be the norm. If the databases are investigator derived, the methods used should be presented. Primary and secondary data resources with full descriptions should be specified (Huston and Naylor 1996). Along with the technical consideration of what to do with missing data, it is also important, from a GIS point of view, to decide whether a sufficient number of records remain for the spatial analysis to produce meaningful results. The true geographic distribution can become distorted when the source of missing values is not identified properly.

Data confidentiality is clearly fundamental to the integrity of small-area analysis. National data collection agencies are aware of the need to preserve confidentiality in connection with or during the process of collection, transmission, storage, use, and dissemination of data (e.g., Medicare beneficiaries). As GIS can be linked to create profiles by mapping individual case information to specific locations, maintaining privacy is a challenging task in health mapping. Confidentiality is of importance for data subjects (i.e., individuals on whom information has been collected), for data suppliers (e.g., clinicians reporting a case), and data users, the research community, who conducts health access and delivery research. In the GIS realm, any maps published would not identify individuals or enable individuals to be identified. Where sensitivity is a problem, groupings or variables are collapsed until aggregated units no longer total small counts. Such methods assure confidentiality, however, may limit the degree of precision required for a given intervention.

Spatial Issues in Rural and Regional Health Disparities

Rural populations without access to comprehensive health services suffer from both short- and long-term effects on their quality of life (Weeks et al. 2004). Access to health care is not uniform across space and access problems are especially pronounced in rural areas (Rosenblatt and Lishner 1991). Although the distances from homes to hospitals vary with geography, people who live in rural areas travel farther than those who live in cities. Distance is an obvious impediment in sparsely populated areas, and by the alarming decline of provider-to-population ratios in rural America (Salsberg and Forte 2002).

In rural or peripheral suburban areas where tracts or zip code areas are large and population tends to concentrate in limited space, physicians often choose to practice at these concentrated areas. Therefore, population-weighted centroids are better than geographic centroids to represent physician locations (Wang and Luo 2005).

The location, hours of operation, and scale of service are important factors in the quality of care, more so in rural settings, because travel times are longer. Consequently, timing of care is very important in rural settings. GIS can support health services planning efforts to make sure that quality health services are available when and where rural residents need them.

For rural areas, utilization data are not always available. Rural health services utilization patterns are mostly retrieved through surveys. Survey questions related to places in which rural residents engage health-related (e.g., number of health-care visits for regular checkup care, chronic care, and acute care) activities are entered into a geographic information system and stratified by county and ethnicity. GIS-assisted analysis determines the importance of spatial behavior as predisposing and enabling factors in rural health-care utilization, controlling for demographic, social, cultural, and health status factors (Arcury et al. 2005).

Smaller rural communities spend more of their health-care resources on avoidable hospital inpatient care than do larger rural communities, leaving smaller rural communities potentially fewer resources to spend on preventive and primary health care. For example, Chen et al. (2009) found that regional variation in ambulatory care–sensitive condition–related hospital charges in the southern United States is generally consistent with the geographic variation in the population's economic status and primary care physician supply.

Health services delivery most commonly involve travel on the part of either the service provider or the patient. The routing models and normative models of facility locations can effectively be integrated into a GIS. Finding the order in which the stops are made that minimizes the total distance traveled is an optimization problem known as the "traveling salesman problem" (Lawler et al. 1985). GIS provides an opportunity to test optimization algorithms using real road networks. In rural areas, where road networks are sparse (i.e., longer distances between residents), home care delivery involves more time to travel between stops. Single provider delivers less service to patients. In this case, the locations of the stops that have to be made are known, and the distances between each pair of stops can be determined by implementing optimization algorithm in a GIS.

Health intervention programs and health policies are increasingly integrating GIS into their health delivery designs, to increase access to and utilization of appropriate preventive and primary health care in rural areas, especially in small and remote communities. Perry and Gesler (2000) used the GPS in the development of a GIS to evaluate the accessibility of primary care services in Andean Bolivia. They were convinced that GIS through field observations with GPS, along with satellite imagery, can help appraise and improve the physical accessibility of health-care services in mountainous areas.

Rural communities that lack resources are less likely to benefit from GIS. Increasing access to quality health care demands that community leaders understand where health inequity and lack of access exist. Social service agencies have been using GIS to promote data sharing, documenting a wide range of conditions, reflecting a broad view of health and to assess the best combination of health and social services to meet rural health needs (Mayor 1999). For example, the Delta Regional Authority (DRA), a federal–state partnership involving eight states in the lower Mississippi Delta region (Illinois, Missouri, Kentucky, Tennessee, Arkansas, Mississippi, Louisiana, and Alabama), addresses the need for better information and data on the health and wellness of DRA communities through GIS (DRA 2015). Most DRA counties and parishes have per capita incomes below the national average, a poverty rate for the region 55% higher than the U.S. national average, infant mortality rates nearly 30% above the national average (Reeder and Calhoun 2002). To make measurable improvements in the health of Delta residents, GIS maps provide data for local leadership to use in benchmarking their health and health-care improvements over time (HDRD 2015). For example, Figure 6.3 shows capacity of DRA hospitals based on number of staffed beds.

Health Profession Distribution

The uneven distribution of health-care providers and patients has been acknowledged as an ongoing issue by the Institute of Medicine (2002). Distance and travel time between the patient and the provider is considered geographic barriers and important spatial factors in assessing health-care access. Significant economic, linguistic, and cultural barriers exist for certain population subgroups, implying the need for consideration of nonspatial

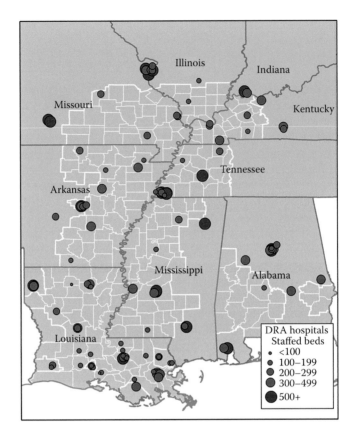

FIGURE 6.3
Capacity of Delta Regional Authority (DRA) hospitals based on the number of staffed beds. Map created by the University of Memphis Spatial Analysis and Geographic Education Laboratory with data obtained from DRA (2015).

factors (i.e., socioeconomic and demographic characteristics) in examining accessibility to primary health care. GIS is the ideal platform to integrate spatial and nonspatial factors and implement methodologies to measure spatial access and identify physician shortage areas.

Accessibility and availability are two important spatial factors to properly understand spatial accessibility (Khan and Bhardwaj 1994; Luo and Wang 2003; Luo 2004). Accessibility is travel impedance between patient and service points and is usually measured in units of distance or travel time. Availability refers to the number of local service locations from which a patient can choose.

Measures of spatial accessibility to health care are classified in four categories: provider-to-population ratios, distance to nearest provider, average distance to a set of providers, and gravitational models of provider influence. All these measures can be conducted quickly and reliably in GIS.

Provider-to-population ratios (i.e., supply ratios of all types of providers, available for download via DAP) are computed within bordered areas such as health service areas, counties, and census units (Schonfeld et al. 1972). The numerator could be the number of physicians or clinics. The denominator is the population size within the area, taken mostly from census or insurance plan enrollment files (e.g., Medicare population). These supply ratios ignore patient border crossing, which can be substantial for small areas. Accessibility within the bordered areas (i.e., the distance dimension) is ignored in supply ratios, assuming patients within the borders have equal access to the physicians.

An alternative method to provider-to-population ratios is the floating catchment method (FCM). Instead of a predefined boundary to compute the provider-to-population ratio, FCM calculates this ratio as a circle of some reasonable radius centered on the census unit (e.g., census tract) centroid and assigns the ratio value to the census tract under consideration. It repeats this process for the rest of the census tracts (i.e., moving the catchment over space) and runs a GIS query to identify the tracts with a ratio less than the standard value (e.g., DHHS standard 1:3500 for primary care) to identify shortage areas. FCM accounts patient–physician interaction across boundaries but still assumes equal access within the census units (Luo 2004).

Distance to the nearest provider is a commonly used measure of spatial accessibility. It is typically measured from a patient's resident or a geometric centroid or a population-weighted centroid of a census unit (Goodman et al. 1992). It is a poor indicator of availability and measured in straight line (Euclidian) distance or estimated travel time via transportation network.

Average distance to a set of providers accounts both accessibility and availability. It is measured from a point of interest, such as a patient residence. From the point of interest, the distance to all providers within a city or a county is summed and averaged, though this approach overweights the influence of providers located near the periphery of the city or the county.

Gravity models assess the potential spatial interaction between any population point and all service points within a reasonable distance (Hansen 1959; Guptill 1975; Guagliardo 2004). Accessibility improves as the number of provider points increases, the capacity at any provider location increases, the distance to provider decreases, or the travel friction decreases.

There are limitations to these spatial accessibility measurement studies. There may be other barriers in the city or defined census units that mileage does not account for, which may make access to certain primary care practice sites more difficult or visits to the EDs more accessible as bus schedules and traffic patterns, which may have influenced care decisions. Future studies may elucidate unidentified barriers, such as bus schedules and traffic patterns, which may have influenced care decisions.

Evaluation of the Effectiveness of Intervention Measures

There is more emphasis now on using GIS to determine and evaluate the effectiveness of intervention measures. GIS analyses have been used by local health departments and health-care providers to support public health intervention activities. For example, Caley (2004) demonstrated that GIS has significant potential to depict community networks to help public health nurses, working together or in transdisciplinary groups, implement population-based interventions at a local level.

One such intervention may be to encourage the insurance companies (e.g., Medicaid HMOs) to utilize a GIS program when assigning a child to a primary care practice site. This small adjustment may reduce geographical barriers to primary care and decrease ED use. By examining the association between ED utilization and geographic variables related to the health-care environment, it may be possible to direct intervention trials toward specific pediatric populations in which barriers to primary care continue to exist.

Medicare Quality Improvement Organization's community collaborative work merits attention for evaluation of the effectiveness of intervention measures. This program uses population-based metrics to assess improvement over time between communities (CMS 2015). Substance Abuse and Mental Health Services Administration (SAMHSA 2015) provides "mapping interventions to different levels of risk" chart showing examples of prevention interventions that can be applied at various stages of the life span.

GIS was used to evaluate the HIV/AIDS programs in Sub-Saharan Africa to assess needs measured in terms of disease prevalence compared with the amount of aid received by each country (Tan 2006). The study concluded that the countries receiving the largest amounts of HIV/AIDS aid were not the neediest, and as much as 75% of the relief funds were for administrative expenses. More GIS-based spatial models can be used longitudinally to monitor changes in primary care access and need, over time as a way to measure the effectiveness of future interventions.

Conclusion

The increasingly important role of health informatics has major consequences. More systems will become automated and integrated. There will be more data available for possible input to GIS. Additional data on patients, programs, facilities, medical diagnostics, treatments, interventions, and their location components will be applicable to GIS processing. In the world of apps, and more maps, online and mobile applications will allow viewing, downloading, and sharing data related to health-care delivery and health-care access.

The outlook for GIS in the health-care access and resource planning appears unlimited. Tomorrow's GIS applications in the health-care industry go beyond managing patient data or analyzing market information. Increasing security, finding ways to release up-to-the-minute information, and providing access to a wider variety of information on the Internet are just a few of the ways GIS changes health care.

References

Agency for Healthcare Research and Quality. (2009). National Healthcare Disparities report, 2008. AHRQ Publication 09-0002. Rockville, MD: U.S. Department of Health and Human Services.

American Hospital Association (AHA). (2000). Essential access—Broadening the safety net. In *Trendwatch*, Washington, DC: American Hospital Association 2(2).

American Hospital Association (AHA). (2008). Annual survey of hospitals and health systems. Chicago, IL: American Hospital Association.

American Hospital Association (AHA). (2001). Research and Trends. http://www.aha.org/index.asp. Accessed October 1, 2001.

American Medical Association (AMA). (1906). Physician master file. http://www.ama-assn.org/ama/pub/about-ama/physician-data-resources/physician-masterfile.page. Accessed October 1, 2015.

Amico, P., Aran, C., and Avila, C. (2010). HIV spending as a share of total health expenditure: An analysis of regional variation in a multi-country study. *PLoS ONE* 5(9): e12997.

Anselin, L. (1992a). SpaceStat tutorial: A workbook for using SpaceStat in the analysis of spatial data. Morgantown, WV: Regional Research Institute, West Virginia University.

Anselin, L. (1992b). Spatial data analysis with GIS: An introduction to application in the social sciences. Technical Report 92-10. Santa Barbara, CA: National Center for Geographic Information and Analysis.

Anthony, D.L., Herndon, M.B., Gallagher, P.M., Barnato, A.E., Bynum, J.P.W., Gottlieb, D.J., Fisher, E.S., and Skinner, J.S. (2009). How much do patients' preferences contribute to resource use? *Health Affairs* 28(3): 864–873.

Arcury, T.A., Gesler, W.M., Preisser, J.S., Sherman, J., Spencer, J., and Perin, J. (2005). The effects of geography and spatial behavior on health care utilization among the residents of a rural region. *Health Services Research* 40: 1.

Barnes, S. and Peck, A. (1994). Mapping the future of health future of health care: GIS applications in health care analysis. *Geographic Information Systems* 4: 30.

Berke, E.M. (2010). GIS: Recognizing the importance of place in primary care research and practice. *The Journal of the American Board of Family Medicine* 23(1): 9–12.

Bernheim, S.M., Grady, J.N., Lin, Z., Wang, Y., Savage, S.V., and Han, L.F. (2010). National patterns of risk-standardized mortality and readmission for acute myocardial infarction and heart failure update on publicly reported outcomes measures based on the 2010 release. *Circulation: Cardiovascular Quality and Outcomes* 3(5): 459–467.

Bernheim, S.M., Horwitz, L., Chohren, P., et al. (2013). Measure updates and speci-fications report: Hospital 30-day readmission following an admission for an acute ischemic stroke (version 2.0). http://www.cms.gov/Medicare/. Accessed March 8, 2013.

Bilheimer, L. and Klein, R.J. (2010). Data and measurement issues in the analysis of health disparities. *Health Services Research* 45(5 Pt 2): 1489–1507.

Bilheimer, L.T., & Sisk, J.E. (2008). Collecting adequate data on racial and ethnic dis-parities in health: The challenges continue. *Health Affairs* 27(2): 383–391.

Billings, J., Georghiou, T., Blunt, I., and Bardsley, M. (2013). Choosing a model to pre-dict hospital admission: An observational study of new variants of predictive models for case finding. *British Medical Journal Open* 3: 8.

Billings, J., Zeitel, L., Lukommic, J., Carey, T.S., Blank, A.E., and Newman, L. (1993). Impact of socioeconomic status on hospital use in New York City. *Health Affairs* 12: 162–173.

Caley, L.M. (2004). Using geographic information systems to design population-based interventions. *Public Health Nursing* 21(6): 547–554.

Carle, A. (2009). Cross-cultural invalidity of alcohol dependence measurement across Hispanics and Caucasians in 2001 and 2002. *Addictive Behaviors* 34: 43–50.

Carret, M.L., Fassa, A.G., and Kawachi, I. (2007). Demand for emergency health service: Factors associated with inappropriate use. *BMC Health Services Research* 7: 131.

Center for Medicare and Medicaid Services (CMS). (2015). Medicare quality improve-ment organizations. https://www.cms.gov/. Accessed December 12, 2015.

Chen, L., Zhang, W., Sun, J., and Mueller, K.J. (2009). The magnitude, variation, and determinants of rural hospital resource utilization associated with hospital-izations due to ambulatory care sensitive conditions. *Journal of Public Health Management and Practice* 15(3): 216–222.

Clarke, T.C., Ward, B.W., Freeman, G., and Schiller, J.S. (2015). Early release of selected estimates based on data from the January–March 2015 National Health Interview Survey. Atlanta, GA: National Center for Health Statistics. http://www.cdc.gov/nchs/nhis.htm. Accessed September 8, 2015.

Cliff, A.D. and Ord, J.K. (1981). *Spatial Processes: Models & Applications*, Vol. 44. London, U.K.: Pion.

Committee on the Future of Emergency Care in the United States Health System. (2007). Hospital-based emergency care: At the breaking point. Washington, DC: The National Academic Press. http://www.nap.edu/catalog/11621/hospital-based-emergency-care-at-the-breaking-point. Accessed October 19, 2015.

Convenient Care Association. Factsheet. http://www.convenientcareassociation.org/CCA%20General%20FACTSHEET.pdf. Accessed August 31, 2008.

Cromley, E. and McLafferty, S. (2002). *GIS and Public Health*. New York: The Guildford Press.

Dartmouth Atlas Project (DAP). (2015). The Dartmouth Institute for Health Policy and Clinical Practice. Hanover, NH: Dartmouth Medical School. http://www.dartmouthatlas.org/.

Davern, M., Davidson, G., Ziegenfuss, J. et al. (2008). Comparing health insurance coverage estimates from four national surveys: Measurement issues and policy implications. Minneapolis, MN: State Health Access Data Assistance Center (SHADAC).

Delta Regional Authority (DRA). (2015). *Today's Delta. A Research Tool for the Region*, 2nd edn. DRA Regional Headquarters. Clarksdale, MS.

Derlet, R.W., Richards, J.R., and Kravitz, R.L. (2001). Frequent overcrowding in US emergency departments. *Academic Emergency Medicine* 8(2): 151–155.

Elliott, M., Finch, B., Klein, D., Ma, S., Do, D., Beckett, M., Orr, N., and Lurie, N. (2008a). Sample designs for measuring the health of small racial/ethnic subgroups. *Statistics in Medicine* 27: 4016–4029.

Elliott, M.N., Fremont, A., Morrison, P.A., Pantoja, P., and Lurie, N. (2008b). A new method for estimating race/ethnicity and associated disparities where administrative records lack self-reported race/ethnicity. *Health Services Research* 43(5, part I): 1722–1736.

Executive Office of the President, Office of Management and Budget. (1997). Revisions to the standards for the classification of federal data on race and ethnicity. http://georgewbushwhitehouse.archives.gov/omb/fedreg/1997standards. html. Accessed December 10, 2015.

Fiscella, K. and Fremont, A. (2006). Use of geocoding and surname analysis to estimate race and ethnicity. *Health Services Research* 41: 1482–1500.

Folland, S. and Stano, M. (1990). Small area variations: A critical review of propositions, methods, and evidence. *Medical Care Review* 47: 419–465.

Gaskin, D.J. and Hoffman, C. (2000). Racial and ethnic differences in preventable hospitalizations across 10 states. *Medical Care Research and Review* 57(Suppl. 1): 85–107.

Gawandi, A. (2011). The hot spotter: Can we lower medical costs by giving the neediest patients better care? Medical Report. *The New Yorker*, pp. 40–51.

Gelfand, A.E., Zhu, L., and Carlin, B.P. (2001). On the change of support problem for spatio-temporal data. *Biostatistics* 2(1): 31–45.

General Accounting Office (GAO). (1995). Health care shortage areas: Designation not a useful tool for directing resources to the underserved (GAO/HEHS-95-2000). Washington, DC: United States General Accounting Office:Health, Education, Human Services Division.

Geronimus, A.T., Bound, J., and Neidert, L.J. (1996). On the validity of using census geocode characteristics to proxy individual socioeconomic characteristics. *Journal of the American Statistical Association* 91(434): 529–537.

Goodman, D.C., Barff, R.A., and Fisher, E.S. (1992). Geographic barriers to cold health services in rural northern New England: 1980 to 1989. *Journal of Rural Health* 8(2): 106–113.

Gornick, M.E., Eggers, P.W., Reilly, T.W., Mentnech, R. M., Fitterman, L.K., Kucken, L.E., and Vladeck, B.C. (2003). Effects of race and income on mortality and use of services among Medicare beneficiaries. *New England Journal of Medicine* 335(11): 791–799.

Green, H. and Himelstein, L. (1998). A cyber revolt in health care. *Business Week* 3600: 154–156.

Grieco, E.M. and Nassidy, R.C. (2001). Overview of race and Hispanic origin. Census 2000 brief. Washington, DC: U.S. Bureau of the Census. http://www.census. gov/prod/2001pubs/cenbr01-1.pdf. Accessed July 10, 2009.

Grumbach, K., Keane, D., and Bindman, A. 1993. Primary care and public emergency department overcrowding. *Am J Public Health.* 83(3): 372–378.

Guagliardo, M.F. (2004). Spatial accessibility of primary care: Concepts, methods and challenges. *International Journal of Health Geographics* 3(1): 3.

Guptill, S.C. (1975). The spatial availability of physicians. *Proceedings of the Association of American Geographers* 7: 80–84.

Hadley, J. and Cunningham, P. (2004). Availability of safety net providers and access to care of uninsured persons. *Health Services Research* 3995: 1527–1546.

Hansen, W.G. (1959). How accessibility shapes land use. *Journal of the American Institute of Planners* 25: 73–76.

Hargraves, J.L. and Hadley, J. (2003). The contribution of insurance coverage and community resources to reducing race disparities in access to care. *Health Services Research* 38(3): 809–829.

Hasnain-Wynia, R. and Baker, D.W. (2006). Obtaining data on patient race, ethnicity, and primary language in health care organizations: Current challenges and proposed solutions. *Health Services Research* 41(4, part I): 1501–1518.

Healthy Delta Research Database (HDRD). (2015). Promoting a Healthy Delta: Healthy Delta Research Database. http://dra.gov/initiatives/promoting-a-healthy-delta/healthy-delta-research-database-healthy-delta-initiative/. Accessed December 10, 2015.

Higgins, P.C. and Taylor, E.F. (2009). Measuring racial and ethnic disparities in health care: Efforts to improve data collection. Policy Brief. Princeton, NJ: Mathematica Policy Research Inc. http://www.mathematica-mpr.com/publications/pdfs/health/disparitieshealthcare.pdf. Accessed July 10, 2009.

Hillenmaier, M., Foster, M., Heinrichs, B., and Heier, B. (2007). Racial differences in parental reports of attention-deficit/hyperactivity disorder behaviors. *Journal of Developmental & Behavioral Pediatrics* 28: 353–361.

Huston, P. and Naylor, D. (1996). Health services research: Reporting on studies using secondary data sources. *Canadian Medical Association Journal* 155(12): 1697–1702.

Institute of Medicine. (2000). *America's Health Care Safety Net: Intact but Endangered.* Washington, DC: National Academy Press.

Institute of Medicine. (2002). *Unequal Treatment: Confronting Racial and Ethnic Disparities in Health Care.* Washington, DC: National Academies Press.

Jankowski, P. and Nyerges, T. (2001). *Geographic Information Systems for Group Decision Making.* London, U.K.: Taylor & Francis Group.

Johnson, P.J., Call, K.T., and Blewett, L.A. (2010). The importance of geographic data aggregation in assessing disparities in American Indian prenatal care. *American Journal of Public Health* 100(1): 122–128.

Joines, J.D., Hertz-Picciotto, I., Carey, T.S., Gesler, W., and Suchindran, C. (2003). A spatial analysis of county-level variation in hospitalization rates for low back problems in North Carolina. *Social Science and Medicine* 56: 2541–2553.

Keenan, T., Rosen, P., Yeates, D. et al. (2007). Time trends and geographical variation in cataract surgery rates in England: Study of surgical workload. *British Journal of Ophthalmology* 91(7): 901–904.

Kershaw S. (2007). Drugstore clinics spread, and scrutiny grows. *New York Times.* http://www.nytimes.com/2007/08/23/nyregion/23clinic.html?_r=1&ref=health&oref=slogin. Accessed October 14, 2007.

Kershaw, S. (2008). Gap in illness rates between rich and poor New Yorkers widens, study says. *New York Times.* http://www.nytimes.com/2007/09/28/nyregion/28gap.html?fta=y. Accessed October 14, 2007.

Khan, A.A. and Bhardwaj, S.M. (1994). Access to health care. A conceptual framework and its relevance to health care planning. *Evaluation and Health Profession* 17(1): 60–76.

Krieger, N. (1992). Overcoming the absence of socioeconomic data in medical records: Validation and application of a census-based methodology. *American Journal of Public Health* 82: 703–710.

Krieger, N. (2005). Painting a truer picture of U.S. socioeconomic and racial/ethnic inequalities: The Public Health Disparities Geocoding Project. *American Journal of Public Health* 95(2): 312–323.

Krieger, N., Chen, J.T., Waterman, D.H., and Subramanian, S.V. (2003a). Race/ethnicity, gender, and monitoring socioeconomic gradients in health: A comparison of area-based socioeconomic measures—The Public Health Geocoding Project. *American Journal of Public Health* 93(10): 1655–1671.

Krieger, N., Chen, J.T., Waterman, P.D., Soabader, M.J., Subramanian, S.V., and Carson, R. (2003b). Choosing area based socioeconomic measures to monitor social inequalities in low birth weight and childhood lead poisoning: The Public health disparities geocoding project. *Journal of Epidemiology and Community Health* 57(3): 186–199.

Krieger, N., Chen, J.T., Waterman, P.D., Soabader, M.J., Subramanian, S.V., and Carson, R. (2003c). Monitoring socioeconomic inequalities in sexually transmitted infections, tuberculosis, and violence: Geocoding and choice of area-based socioeconomic measures-The Public health Disparities Geocoding Project (US). *Public Health Reports* 118(3): 240–260.

Lambe, S., Washington, D.L., Fink, A., Herbst, K., Liu, H., Fosse, J.S., and Asch, S.M. (2002). Trends in the use and capacity of California's emergency departments, 1990–1999. *Annals of Emergency Medicine* 39(4): 389–396.

Lang, L. (2000). *GIS for Health Organizations*. Redlands, CA: ESRI Press.

Lawler, E.L., Lonsa'a, L.K., Rinnooy Kan, A.G.H., and Shimoys, D.B. (1985). *The Traveling Salesman Problem*. New York: John Wiley & Sons.

Lee, R.C. (1991). Current approaches to shortage areas designation. *Journal of Rural Health* 7: 437–450.

Leung, B., Luo, N., So, L., and Quan, H. (2007). Comparing three measures of health status (perceived health with Likert-Type scale, EQ-5D, and number of chronic conditions) in Chinese and White Canadians. *Medical Care* 45: 610–617.

Lucas, F.L., Sirovich, B.E., Gallagher, P.M., Siewers, A.E., and Wennberg, D.E. (2010). Variation in cardiologists' propensity to test and treat: Is it associated with regional variation in utilization? *Circulation: Cardiovascular Quality and Outcomes* 3(3): 253–260.

Ludwig, A., Fu, R., Warden, C., and Lowe, R.A. (2009). Distances to emergency department and to primary care provider's office affect emergency department use in children. Portland, OR: The Society of American Emergency Medicine, Western Regional Research Forum.

Luo, W. (2004). Using a GIS-based floating catchment method to assess areas with shortage of physicians. *Health and Place* 10: 1–11.

Luo, W. and Wang, F. (2003). Measures of spatial accessibility to health care in a GIS environment: Synthesis and a case study in Chicago region. *Environment and Planning B* 30(6): 865–884.

Magner, D., Mirocha, J., and Gewertz, B.L. (2009). Regional variation in the utilization of carotid endarterectomy. *Journal of Vascular Surgery* 49(4): 893–901; discussion 901.

Maheswaran, R. and Maheswaran, R. (1997). Supply of in-patient medical services for elderly people and geographical variation in medical admissions in a health district in England. *Public Health* 111(6): 411–415.

Majeed, A., Eliahoo, J., Bardsley, M., Morgan, D., and Bindman, A.B. (2002). Variation in coronary artery bypass grafting, angioplasty, cataract surgery, and hip replacement rates among primary care groups in London: Association with population and practice characteristics. *Journal of Public Health Medicine* 24(1): 21–26.

Makarov, D.V., Loeb, S., Landman, A.B., Nielsen, M.E., Gross, C.P., Leslie, D.L., Penson, D.F., and Desai, R.A. (2010). Regional variation in total cost per radical prostatectomy in the healthcare cost and utilization project nationwide inpatient sample database. *Journal of Urology* 183(4): 1504–1509.

Martin, S., Smith, P., Martin, S., and Smith, P. (1996). Explaining variations in inpatient length of stay in the National Health Service. *Journal of Health Economics* 15(3): 279–304.

Matlock, D.D., Peterson, P.N., Sirovich, B.E., Wennberg, D.E., Gallagher, P.M., and Lucas, F.L. (2010). Regional variations in palliative care: Do cardiologists follow guidelines? *Journal of Palliative Medicine* 13(11): 1315–1319.

Mayor, T. (1999). Pathways connects Atlanta's homeless, poor to services. *Civic.com* 3(12): 18–19.

McMahon Jr., L.F., Wolfe, R.A., Griffith, J.R., and Cuthbertson, D. (1993). Socioeconomic influence on small area hospital utilization. *Medical Care* 31(Suppl.): YS29–YS36.

Meade, M. and Erickson, R. (2000). *Medical Geography*. New York: Guildford Press.

Mehrotra, A., Wang, M.C., Lave, J.R., Adams, J.L., and McGlynn, E.A. (2008). Retail clinics, primary care physicians, and emergency departments: A comparison of patients' visits. *Health Affairs* 27(5): 1272–1282.

Mulholland, C., Harding, N., Bradley, S. et al. (1996). Regional variations in the utilization rate of vaginal and abdominal hysterectomies in the United Kingdom. *Journal of Public Health Medicine* 18(4): 400–405.

Mulley, A.G. (2009). Inconvenient truths about supplier induced demand and unwarranted variation in medical practice. *British Medical Journal* 339: 1007–1009.

O'Hare, A.M., Rodriguez, R.A., Hailpern, S.M., Larson, E.B., and Kurella Tamura, M. (2010). Regional variation in health care intensity and treatment practices for end-stage renal disease in older adults. *The Journal of the American Medical Association* 304(2): 180–186.

Openshaw, S. (1984). The modifiable areal unit problem. Concepts and techniques in modern geography, No. 38. Norwich, U.K.: Geo Books.

Perry, B. and Gesler, W. (2000). Physical access to primary health care in Andean Bolivia. *Social Science and Medicine* 50: 1177.

Pollack, C.E. and Armstrong, K. (2009). The geographic accessibility of retail clinics for underserved populations. *Archives of Internal Medicine* 169(10): 945–949.

Porter, M.E. (2010). What is value in health care? *New England Journal of Medicine* 363(26): 2477–2481.

Prewitt, K. (2005). Racial classification in America: Where do we go from here? *Daedalus* 134: 5–17.

Radke, J. and Mu, L. (2000). Spatial decomposition, modeling and mapping service regions to predict access to social programs. *Geographic Information Science* 6: 105–112.

Rafalski, E. and Zun, L. (2004). Using GIS to monitor emergency room use in a large urban hospital in Chicago. *Journal of Medical Systems* 28(3): 311–319.

Reeder, R. and Calhoun, S. (2002). Federal funding in the delta. *Rural America* 17(4): 20–30.

Renger, R., Cimetta, A., Pettygrove, S., and Rogan, S. (2002). Geographic information systems (GIS) as an evaluation tool. *The American Journal of Evaluation* 23(4): 469–479.

Retail-Based Clinic Policy Work Group, AAP. (2006). AAP principles concerning retail based clinics. *Pediatrics* 118(6): 2561–2562.

Rosenblatt, R.A. and Lishner, D.M. (1991). Surplus or shortage? Unraveling the physician supply conundrum. *The Western Journal of Medicine* 154: 43–50.

Rust, G., Ye, J., Baltrus, P., Daniels, E., Adesunloye, B., and Fryer, G.E. (2008). Practical barriers to timely primary care access: Impact on adult use of emergency department services. *Archives of Internal Medicine* 168(15): 1705–1710.

Salsberg, E.S. and Forte, G.J. (2002). Trends in the physician workforce, 1980–2000. *Health Affairs* 21(5): 165–173.

Schonfeld, H.K., Heston, J.F., and Falk, I.S. (1972). Numbers of physicians required for primary medical care. *New England Journal of Medicine* 286(11): 571–576.

Shen, Y.C. and Hsia, R.Y. (2010). Changes in emergency department access between 2001 and 2005 among general and vulnerable populations. *American Journal of Public Health* 100(8): 1462–1469.

Steele, C.B., Cardinez, C.J., Richardson, L.C., Tom-Orme, L., and Shaw, K.M. (2008). Surveillance for health behaviors of American Indians and Alaska Natives—Findings from the behavioral risk factor surveillance system, 2000–2006. *Cancer* 113(5 Suppl.): 1131–1141.

Stukel, T.A., Fisher, E.S., Wennberg, D.E. et al. (2007). Analysis of observational studies in the presence of treatment selection bias: effects of invasive cardiac management on AMI survival using propensity score and instrumental variable methods. *The Journal of American Medical Association* 297(3): 278–285.

Stukel, T.A., Lucas, F.L., and Wennberg, D.E. (2005). Long-term outcomes of regional variations in intensity of invasive vs. medical management of medicare patients with acute myocardial infarction. *The Journal of American Medical Association* 293(11): 1329–1337.

Substance Abuse and Mental Health Services Administration (SAMHSA). (2015). Mapping interventions to different levels of risk. http://www.samhsa.gov/. Accessed December 8, 2015.

Suruda, A., Burns, T.J., Knight, S., and Dean, J.M. (2005). Health insurance, neighborhood income, and emergency department usage by Utah children 1996–1998. *BMC Health Services Research* 5(1): 29.

Sutterhealth. (2015). Community benefit and our mission. http://www.sutterhealth. org/communitybenefit/mission.html. Accessed December 11, 2015.

Tan, X. (2006). Evaluating HIV/AIDS Programs. *ArcUser Magazine,* July-September 2006. Redlands, CA: ESRI Press.

Trendwatch. (2000). Essential access—Broadening the safety net. *American Hospital Association* 2(2): 1–8.

Tu, H.T. and Cohen, G.R. (2008). *Checking Up on Retail-Based Health Clinics: Is the Boom Ending?* New York: Commonwealth Fund.

Tufte, E.R. (2001). *The Visual Display of Quantitative Information,* 2nd edn. Cheshire, England: Graphics Pr.

U.S. Bureau of Labor Statistics. (2006). The current population survey (CPS). Washington, DC: Bureau of Labor Statistics.

U.S. Department of Health and Human Services. (2000a). Healthy people 2010: Understanding and improving health. Washington, DC.

U.S. Department of Health and Human Services. (2000b). Health resources and services administration. http://datawarehouse.hrsa.gov/tools/dataportal. aspx. Accessed October 6, 2015.

U.S. Department of Health and Human Services (DHHS). (2004). Website maintained by the health resources and services administration. http://bphc.hrsa.gov/dsd. Accessed October 10, 2015.

U.S. Department of Health and Human Services (DHHS). (2010). Bureau of Health Professions. Health Workforce Profiles, CA.

Vanasse, A., Courteau, J., and Courteau, M. (2012). The interactive atlas on health inequalities. *Spatial and spatio-temporal epidemiology* 3(2): 129–140.

Vinig, B.A. and McBean, M. (2001). Administrative data for public health surveillance and planning. *Annual Review of Public Health* 22: 213–230.

Walls, C.A., Rhodes, K.V., and Kennedy, J.J. (2002). The emergency department as usual source of medical care: Estimates from the 1998 National Health Interview Survey. *Academic Emergency Medicine* 9(11): 1140–1145.

Wang, F. and Luo, W. (2005). Assessing spatial and nonspatial factors for healthcare access: Towards an integrated approach to defining health professional shortage areas. *Health and Place* 11: 131–146.

Weber, E.J., Showstack, J.A., Hunt, K.A., Colby, D.C., and Callaham, M.L. (2005). Does lack of a usual source of care or health insurance increase the likelihood of an emergency department visit? Results of a national population-based study. *Annals of Emergency Medicine* 45(1): 4–12.

Weeks, W.B., Kazis, L.E., Shen, Y., Cong, Z., Ren, X.S., Miller, D., Lee, A., and Perlin, J.B. (2004). Differences in health-related quality of life in rural and urban veterans. *American Journal of Public Health* 94(10): 1762–1767.

Weinick, R.M., Caglia, J.M., Friedman, E., and Flaherty, K. (2007). Measuring racial and ethnic health care disparities in Massachusetts. *Health Affairs* 26(5): 1293–1302.

Weinick, R.M., Zuvekas, S.H., and Cohen, J.W. (2000). Racial and ethnic difference in access to and use of health care services, 1977 to 1996. *Medical Care Research and Review* 57(Suppl. 1): 36–54.

Welch, H.G., Sharp, S.M., Gottlieb, D.J., Skinner, J.S., and Wennberg, J.E. (2011). Geographic variation in diagnosis frequency and risk of death among medicare beneficiaries. *The Journal of American Medical Association* 305(11): 1113–1118.

Wennberg, J.E. (1984). Dealing with medical practice variations. A proposal for action. *Health Affairs* 3: 6–32.

Wennberg, J.E. (1987). Population illness rates do not explain population hospitalization rates. *Medical Care* 25: 354–359.

Wilner, D. (1977). The role of the emergency department in the delivery of rural primary care. *The Journal of the Maine Medical Association* 68(11): 401–408.

Zuvekas, S.H. and Weinick, R.M. (1999). Changes in access to care, 1977–1996: The role of health insurance. *Health Services Research* 34(1): 271–280.

Conclusion

Development and Evolution of GIS in the Context of Health Inequalities

Before reading this book, if you had been aimlessly wondering of the following questions, then hopefully you found this book informative, and it should put you on the road to explore the many faces of geographic information systems (GIS) and spatial data analysis in health inequalities research:

- Who are subject to significant health inequalities?
- What is it about certain places that have such stark differences in particular health outcomes?
- When do adverse health events occur?
- Where are more interventions needed to be done?
- Why do some populations not have equal chance to achieve their best health?
- Which spatial measures do I use to assess health inequalities?
- How do I analyze health inequalities using GIS?

This book has demonstrated the use of GIS in health inequalities. The title of this book *Spatial Health Inequalities: Adapting GIS Tools and Data Analysis* comes from the quest to investigate the belief "who you are depends upon where you are" and the fact that health inequality is both a people and a place concern. This book aims to expand our understanding of inequalities for adverse health outcomes in a geographic context. A set of health outcomes (e.g., chronic diseases, infectious diseases), representative of populations experiencing health inequalities, were discussed. The examples in this book illustrated key GIS applications and issues in identifying health inequalities. The emphasis was on GIS operations and applications rather than on statistical issues.

Health inequalities remain a persistent problem globally, leading to certain groups being at higher risk of being uninsured, having more limited access to care, experiencing poorer quality of care, and ultimately experiencing worse health outcomes. While health inequalities are commonly viewed through the lens of race and ethnicity, this book demonstrated that they occur across a broad range of dimensions and reflect a complex set of individual, social, and environmental factors. GIS has been utilized to incorporate socioeconomic and neighborhood information into the risk assessment process and enable health officials to evaluate the racial/ethnic, socioeconomic, and environmental characteristics of those at highest risk (Simons and Van Derslice 1995).

As global health burden differs significantly by geography, race/ethnicity, and socioeconomic status, addressing health inequalities spatially is emerging as a critical strategy for the health community. It is a burgeoning area for the use of GIS. With GIS, researchers can identify disparities and problem areas and also seek to reduce the burden of health inequalities by studying interventions, their impacts in defined populations, and the means by which they can be better used (Pickle et al. 2006).

The multilevel and multifactorial nature of health inequalities are best addressed with GIS, which also accommodates the multifaceted nature of space, spatial representations, and the societal implications of geospatial information. GIS functions as a true integrated multidisciplinary technology. It uses data and techniques from many professions and academic disciplines, and its applications are found in a diverse set of fields (e.g., geography, cartography, statistics, computer science, and subject-specific fields). Recognizing that most of these fields are relevant for the study of health inequalities, the most efficient progress on eliminating health inequalities can be achieved by collaboration and consensus development on data sharing and confidentiality (i.e., in connection with or during the process of collection, transmission, storage, use, and dissemination of data) across international, federal agencies and among governments, academia, disease registries (e.g., cancer, diabetes), and other interested groups. It is generally accepted that any items of demographic (e.g., race/ethnicity) or health data that would permit the identification of an individual person are confidential. However, in certain circumstances (i.e., small area mapping), other types of data may enable individuals to be identified, particularly if they have a rare disease (e.g., cancer) and live in a sparsely populated area (Quinn 1992). For preserving the rights and confidentiality of individuals, the data are expected by the data providers to be as disaggregated as possible. Researchers then assign geographic codes to small subunits of data to facilitate analysis of the spatial relationships among variables. The ability to map these variables is an important feature of GIS, since visually displaying these interrelationships can lead to etiologic and societal clues.

Institutional issues and legal and ethical considerations will continue to redefine public access to confidential health information. Future research needs to evaluate methods and resource requirements for the health information in a way that both protects confidentiality and maximizes availability. Methods evaluated should include external release of de-identified data and internal geographic analysis of the data through software agents capable of analyzing the original data and returning results that do not include any identifiable information (Boulos et al. 2006). This can lead to more effective and appropriate decision-making for health inequalities and their interventions.

Technological changes continue to drive the evolution of GIS, making it difficult to close the digital divide that limits access (i.e., differential access to hardware, software, and technical training) to computer-based communication and technological advances in poor communities, countries, and remote/rural areas. Health communities in even most technologically advanced countries have only recently embraced the use of GIS methods. We are indeed at a key point in which the technological issues (i.e., digital divide) need to be addressed before GIS methods can be fully used to lessen the health burden of communities subject to significant health inequalities. An international collaborative effort between practitioners, national public health agencies, researchers based in institutions of higher education, non-profit organizations, for-profit health-care providers, software companies, and community groups needs to address technical issues in GIS implementation and work as a unifying source toward how, when, and where GIS can and will be used given their capabilities, resources, and political power.

Data will continually drive GIS implementation in health settings, as more and more digital spatial health data become available. In a deeply interconnected world of smartphones, tablets, and Google searches, where maps are replaced by apps, pairing Big Data (i.e., massive amounts of information interpreted by analytics to provide trends or patterns) with GIS is one path to create more complete patient profiles for better population health and to pinpoint populations who are at risk for health inequalities (Chen et al. 2006; Bedard 2014). Through analysis of vast data streamed from social media, multiple devices, and international statistics, researchers can better predict more precise locations of the next outbreaks and potentially control future diseases. Indeed, this is a critical time for Big Data in addressing health inequalities and for consideration of GIS and ways of thinking to fully incorporate Big Data ideas into intervention and prevention efforts. GIS tools for Big Data processing can facilitate predictive modeling for infectious disease spread, policy making in health care, crime detection, disaster response, and so on. GIS output maps can integrate unstructured data (e-mails, blogs, social media content, sensor data [e.g., reservoir host animal], meteorological data, driving times, etc.), in some cases in real time. This is useful for exploring infectious disease epidemiology, targeting early warning signs, and improving patient safety, crisis/disaster management, health-care planning, and population health management strategies. For future research, the convergence of Big Data and GIS leads to deeper insights in health inequalities, efficiency in health care (e.g., fewer readmissions), improved patient and health professional engagement, and better population health.

I hope that I have conveyed the amazing breadth of GIS use in studying health inequalities. While I have focused somewhat narrowly on several representative health outcomes, there are so many others that can be explored with GIS under the realm of health inequalities, in addition to the examples discussed in this book. The recommendations that emerged from this book will be an important guiding force in helping advance a GIS and health inequalities research agenda and ultimately should help reduce future health inequalities. As these topics are applicable beyond the United States, the results of this book can serve as a valuable resource for setting research agendas around the world.

References

Bedard, Y. (2014). Beyond GIS: Spatial on-line analytical processing and big data [Online]. Orono, ME: University of Maine.

Boulos, K.M.N., Cai, Q., Padget, J.A., and Rushton, G. (2006). Using software agents to preserve individual health data confidentiality in micro-scale geographical analyses. *Journal of Biomedical Informatics* 39: 160–170.

Chen, Y., Suel, T., and Markowetz, A. (2006). Efficient query processing in geographic web search engines. *Proceedings of the 2006 Association for Computing Machinery (ACM) SIGMOD International Conference on Management of Data*, New York, NY: pp. 277–288 http://dl.acm.org/citation.cfm?id=1142505. Accessed October 5, 2015.

Pickle, L.W., Szczur, M., Lewis, D.R., and Stinchcomb, D.G. (2006). The crossroads of GIS and health information: A workshop on developing a research agenda to improve cancer control. *International Journal of Health Geographics* 5: 51.

Quinn, M.J. (1992). Confidentiality. In D. English (Ed.), *Confidentiality in Geographical and Environmental Epidemiology: Methods for Small-Area Studies*. Oxford, U.K.: Oxford University Press, pp. 132–140.

Simons, V. and VanDerslice, J. (1995). The use of GIS for incorporating environmental equity concerns into the risk assessment process. Abstracts of the *International Symposium on Computer Mapping in Epidemiology and Environmental Health*, Tampa, FL, February 1995, p. 60.

Index